Key Concepts STEM

SCIENCE TECHNOLOGY ENGINEERING MATH

PHYSICS

MATTER, ENERGY, AND HEAT

General Editor: Anne O'Daly

BROWN BEAR BOOKS

Published by Brown Bear Books Ltd
4877 N. Circulo Bujia, Tucson, AZ 85718
USA
and
Studio G14, Regent Studios, 1 Thane Villas,
London N7 7PH, UK

Cataloging-in-Publication data available upon request.

ISBN 978-1-83572-081-3 (ALB)
ISBN 978-1-83572-086-8 (paperback)
ISBN 978-1-83572-091-2 (ebook)

Children's Publisher: Anne O'Daly
Design Manager: Keith Davis
Picture Manager: Sophie Mortimer
Consultant: Chris Woodford

Manufactured in the United States of America
CPSIA compliance information: Batch#AG/5665

Picture Credits
Front Cover: Shutterstock: Jag cz.
Interior: Dreamstime: Stockvectorwin 4-5; **iStock**: AntonVimages 9, dallaspapaazzo 6-7, Gilltukha 20-21, studiocasper 13, ventdusud 12; **Photos.com**: 41; **Public Domain**: Maarten Visser 10-11; **Shutterstock**: ABCDstock 24-25, Can Balcioglu 26, Willyam Bradbury 38-39, Michael Coddington 36-37, Anton Gozdikov 56-57, In Green 19, johnmoorefour 1, 33, Pavel Kilmenko 48-49, Kingcraft 45, Photographee.eu 54, photos manYmage 43, E. G. Pors 16, Philip Rozenski 60-61, Slipitsskaya 46, Thoron 17, VisionDive 58-59, Victor Zastoloskiy 28-29; **United States Government**: Lt Col. William Thurmond, Photo Courtesy of U.S. Army 23.

STEM

Key Concepts in STEM: Physics describes the key theories in the study of physics and their practical applications. *Matter, Energy, and Heat* introduces the states of matter (solid, liquid, or gas) found in the world around us. Explanatory diagrams and informative photographs, detailed features on related aspects of the topics covered and the main scientists involved in the advancement of physics, and definitions of key "Science Words" all enhance the coverage. "Try This" sidebars outline experiments that learners can follow as a first step to practical investigations.

Publisher's note to educators and parents: Our editors have carefully reviewed the web sites included at the end of this book to ensure that they are suitable for students. Many web sites change frequently, however, and we cannot guarantee that a site's future contents will continue to meet our high standards of quality and educational value. Be advised that students should be closely supervised whenever they access the Internet.

Contents

ATOMS AND MOLECULES

Everything in the world is made from tiny particles. The smallest particles that can exist on their own are called atoms. Often, atoms join together to form larger particles called molecules. All chemical compounds are made up of molecules.

There are just over 100 different chemical elements, each with its own kinds of atom. Most elements are metals, such as iron, copper, and aluminum. A piece of iron is made up of trillions and trillions of iron atoms. Some elements are gases, such as oxygen and hydrogen. Only two elements are liquid at ordinary temperatures. They are the silvery liquid metal mercury and a poisonous reddish brown liquid that is called bromine in the English-speaking world.

The characteristics of substances formed when atoms of different elements combine are often very different from those of the individual elements. For example, when the

This is a model of an atom, showing the central core, or nucleus, and electron shells.

CHEMICAL BONDING

In ionic bonding (top), one atom gives one or more electrons to another atom to form a chemical compound that consists of ions. Here, sodium and chlorine combine to make sodium chloride. In covalent bonding (bottom), the atoms share electrons. Here, an oxygen atom shares electrons with two hydrogen atoms to form a molecule of water.

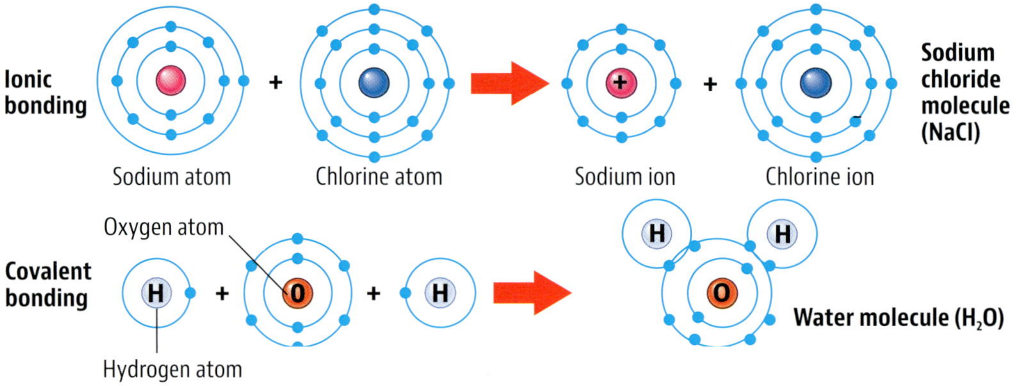

DIFFERENT KINDS OF CARBON

Diamond, graphite, and fullerenes are all forms of carbon. Graphite is solid black, shiny, and relatively soft. It has layers of carbon atoms that slide easily over one another. Diamonds are molecules of carbon atoms held together in a rigid structure that makes them extremely hard. The C_{60} fullerene, or "buckyball", is a sphere of 60 carbon atoms. Its polygonal structure makes it strong. Other fullerenes include carbon nanotubes, which are used in biotechnology and medicine.

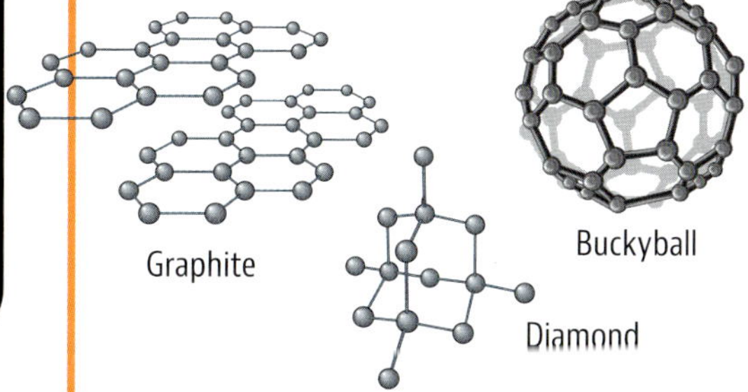

metal copper combines with the gas oxygen, it forms molecules of the nonmetal copper oxide. And the two gases hydrogen and oxygen combine to form the liquid we know as water.

Atoms are not solid balls of matter. They consist of a central core, or nucleus, surrounded by one or more electrons. Electrons are therefore even smaller than the smallest atom—about one two-thousandth the size. Each electron carries a single negative electrical charge. The nucleus of the atom is also charged. Nuclei have a positive charge that balances the negative charges of the electrons. The electrons in an atom are arranged around the nucleus in layers called shells, like the different layers inside an onion. The joining of atoms to form molecules involves the electrons. Some kinds of atom—chiefly metals—can lose one or more electrons. This leaves them with a positive charge. Charged atoms are called ions. Other kinds of atom—mainly nonmetals—can gain one or more electrons. This gives them a negative charge. They become negative ions. If a metal such as sodium reacts with a nonmetal such as chlorine, each sodium atom gives an electron to a chlorine atom. They combine to form a molecule of sodium chloride, or table salt.

SCIENCE WORDS

Atom: The smallest part of a chemical element that can exist on its own. It has a central nucleus (made up of protons and neutrons), surrounded by electrons.

Molecule: A combination of at least two atoms that forms the smallest unit of a chemical element or compound.

BEHAVIOR OF GASES

Like all kinds of matter, a gas is made up from atoms or molecules. But these particles do not stand still. They rush around, hitting one another and the walls of their container. As they collide with the container, gas pressure rises.

Gases make up one of the three everyday states of matter—the other two are liquids and solids. We are surrounded by gas all the time because the air we breathe is a gas. Air is a mixture of gases, mostly nitrogen and oxygen. You cannot see air, yet it has mass, and it exerts a pressure on everything in contact with it. The air in an average room weighs about 175 lb (about 80 kg). The weight of all the air in the atmosphere totals around 5,000 million million tons. At sea level, the pressure is about 14 lb per square inch of surface (about 1 kg per sq. cm). The pressure of the atmosphere can be measured using a barometer, of which there are various kinds.

The Italian scientist Evangelista Torricelli (1608–1647) invented the first mercury barometer in 1644. He wanted to prove that vacuums are caused by atmospheric pressure. He took a long glass tube closed at one end and filled it with mercury. The tube was turned upside down and the open end was lowered into a bowl of mercury. The level of the mercury in the tube fell slightly, then it stopped. The rest of the mercury remained in the tube. Torricelli reasoned that it was the atmosphere pressing on the surface of the mercury in the bowl that held up the column of mercury. The space above the mercury in the closed end of the tube had no air at all.

Predicting the Weather

Normal atmospheric pressure will hold up a column of mercury about 30 in (76 cm) tall. Because atmospheric pressure varies from day to day depending on the weather, the height of the mercury column also varies. People soon learned how to forecast the weather by watching the changes in pressure as revealed by a barometer. Clear, dry weather, for example, can usually be expected when the atmospheric pressure is on the increase. When the atmospheric pressure is falling, this generally indicates rainy weather. Atmospheric pressure also decreases with height above the ground. It falls about 0.4 in (about 1 cm) for every 328-ft (100-meter) increase in height. Barometers can therefore be used for measuring altitude.

Many of the altimeters carried by aircraft are sensitive barometers. They are usually the aneroid type ("aneroid" comes from Greek words meaning "no liquid"). An aneroid barometer does not contain mercury. The heart of the instrument is a closed chamber containing no air. As the external air pressure varies, the small vacuum chamber changes shape, operating a system of levers to move a pointer that indicates the value of the pressure on a dial. In an altimeter, the dial is calibrated in height, not in pressure.

SCIENCE WORDS

Atmospheric pressure: The pressure of the Earth's atmosphere at any point on its surface (caused by the weight of the column of air above it). Atmospheric pressure decreases with altitude (height above ground).

Boyle's Law: At constant temperature the pressure of a gas is inversely proportional to its volume. For example, if the pressure increases, then the volume decreases.

Gas: A state of matter in which the molecules move at random. A gas in a container takes on the size and shape of the container.

Pressure: The amount of force pressing on a particular area.

Helium is one of the two gases that are less dense than air—the other is hydrogen. Unlike hydrogen, helium does not burn. It is used to fill weather balloons and airships.

Other Kinds of Gas

Not all gases are as dense as air. Two light gases are hydrogen and helium. Hydrogen is dangerously flammable. It was once used for filling balloons and airships, but after it caused several disasters, the most famous being the explosion of the German airship *Hindenburg* in 1937, its use was abandoned. Today, airships are filled with helium, which being less dense than air causes airships to float upward.

Other gases are used as fuels. Methane occurs underground in natural gas. The similar gases propane and butane are flammable but nontoxic gases obtained from crude oil in an oil refinery. They are sold as bottled gas for heating and lighting in campers and trailers. The same gases are also known as LPG (liquefied petroleum gas) and are being used as a cleaner alternative to gasoline for automobile engines.

Acetylene is a fuel gas that is burned in oxygen in an oxyacetylene torch. The very hot flame is used for welding and cutting steel and other metals. The acetylene gas is liquefied and supplied in steel cylinders. Carbon dioxide is a very dense gas composed of one carbon and two oxygen atoms. In its solid state it is called dry ice. Nothing will burn in carbon dioxide, which is why it is used in fire extinguishers.

Properties of Gases

A gas that is confined within a closed container exerts a pressure. That is because the rapidly moving molecules of gas collide with the walls of the container. It is also why we have to keep the container closed—a gas will soon escape from an open container. If we decrease the volume of gas by making the container smaller, the pressure of the gas increases. For example, the pressure doubles if the size of the container is halved. This relationship between gas pressure and volume was first discovered by the British scientist Robert Boyle (1627–1691), and for this reason it is known as Boyle's Law.

BOYLE'S LAW

These diagrams show the effect of increasing the pressure on a gas. When the pressure (P) is increased from 1 bar to 2 bars, the volume (V) is halved from 8 cu m to 4 cu m. Increasing the pressure to 4 bars reduces the volume to only 2 cu m. In other words, pressure is inversely proportional to volume. Put another way, the product of pressure and volume is constant.

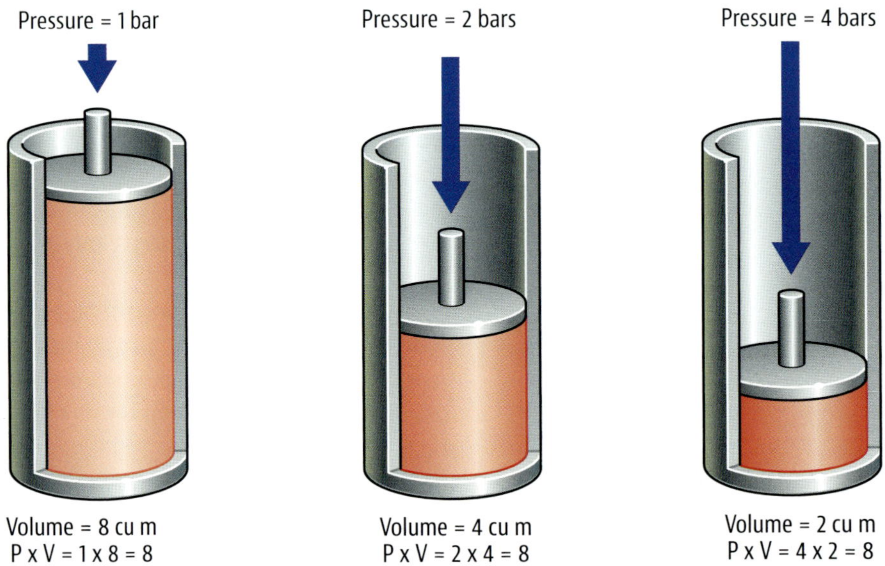

An oxyacetylene welder at work. Acetylene gas burns in oxygen with an extremely hot flame, which can be used for cutting or welding metal. Welders wear dark glasses to protect their eyes from the bright light.

Heating a gas also increases the pressure as long as the volume is kept the same. That is because the hotter gas molecules move faster and collide with the walls more often. If the volume is not kept constant, heating a gas causes it to expand and take up more room. An expanding gas can be made to do work. Machines, such as the steam engine and the jackhammer, make use of this property.

A flow of gas can also be made to do work. The earliest device to make use of moving gas was the sailing ship, where the moving gas is the wind. Later, people harnessed the wind with windmills. Many windmills had sails that resemble those on a sailing ship. Windmills like this can still be seen in Mediterranean countries such as Spain and Greece.

ROBERT BOYLE

Robert Boyle was born in Munster, Ireland, in 1627. He was the son of the Earl of Cork. After studying in Switzerland he returned to England in 1644, and settled in Oxford 10 years later. Boyle carried out many experiments in both physics and chemistry. He investigated electricity, crystals, and the behavior of gases. He also invented an air pump and studied the effects of pressure on gases. In 1662, he formulated Boyle's Law, which states that at constant temperature, the pressure of a gas is inversely proportional to its volume. He later moved to London and became a founder member of the Royal Society in 1662. Boyle died in 1691.

MERCURY MANOMETER

A simple device for measuring gas pressures is called a manometer. This consists of a U-shaped glass tube containing mercury. One end of the tube is open to the air, and the other end is connected to the gas supply. The pressure of the gas forces mercury up the open arm of the manometer. The difference in heights of the mercury columns is a measure of the pressure.

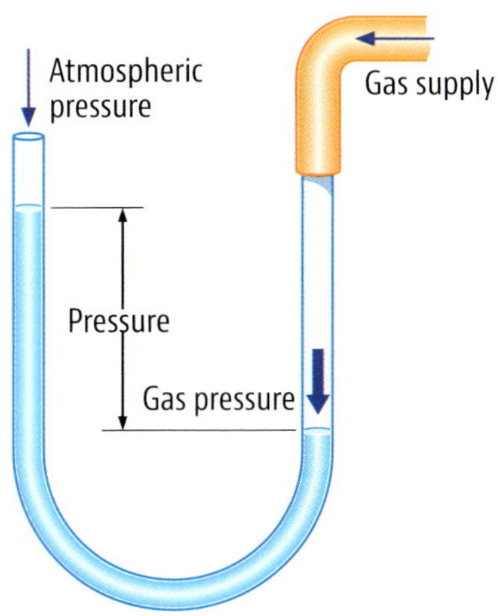

BEHAVIOR OF LIQUIDS

Liquids are the "middle" state of matter. They are denser than gases, but not as dense as solids. Like gases, liquids flow easily and have to be kept in a container.

The molecules that make up a liquid are free to move around. That is why a liquid has no definite shape—it just takes on the shape of its container. The container does not need a lid to prevent the liquid from escaping, unlike a gas. Also unlike a gas, a liquid cannot be compressed: increasing the pressure on a liquid does not make it occupy a smaller volume. But a liquid does exert pressure on the sides and bottom of its container, and on anything that happens to be in it. This pressure depends on the liquid's density and depth. The greater the depth of a liquid, the greater the corresponding pressure.

Another property of a liquid is called viscosity, which is a measure of how thick it is. Machine oil and molasses are viscous liquids, which flow only slowly. Water and alcohol are less viscous, and they flow easily. This is because the molecules in water, for example, slide past one another much more easily than do the molecules in molasses.

CAPILLARY ACTION

The shape of the meniscus in a capillary tube depends on the density of the liquid and its ability to wet the glass. Water has a concave (curved downward) meniscus; mercury has a convex (curved upward) meniscus.

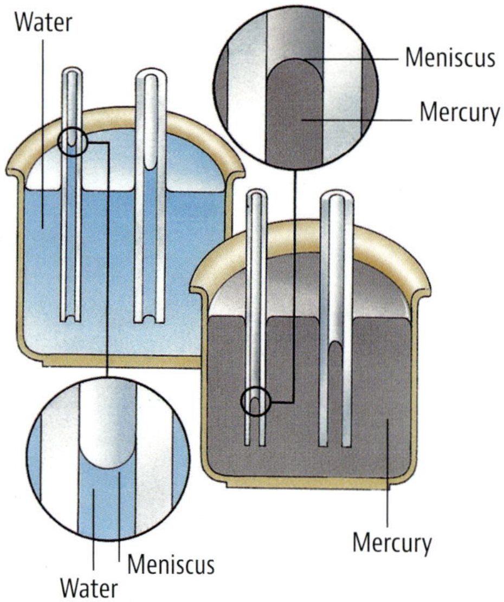

Surface Tension and Surfactants

The molecules at the surface of a liquid attract one another. This causes an effect called surface tension, which makes the liquid behave as if it has an elastic "skin" on its surface. This makes it possible to float a needle on the surface of water in a glass and allows bugs, such as pond skaters, to walk across a pond. It is also why bubbles and liquid droplets are spherical.

Water is the most common liquid on Earth. Here a "water bomber" aircraft drops water on a forest fire in an attempt to put it out.

A substance called a surfactant, also known as a wetting agent, can reduce surface tension. Detergents are surfactants, and adding them to washing water allows the water to push into small crevices in closely woven cloth and pry away any dirt that may be lodged there.

Capillarity

Another effect of surface tension is called capillary action, or capillarity. It is this effect that makes blotting paper or a sponge soak up water. It can be observed by lowering a narrow glass tube vertically into a liquid. With a liquid such as water, the liquid rises up the tube a little way. If you look closely at the surface of the liquid inside the tube, you will see that it is curved—it is higher at the sides of the tube than at the center. This curve, called a meniscus, is saucer-shaped in the case of water. But in a dense liquid such as mercury, it is curved upward—higher at the center than at the sides.

SCIENCE WORDS

Capillarity: Also called capillary action, the movement of a liquid up or down a narrow tube, caused by the attraction between its molecules and those of the tube. The surface of the liquid is curved into a meniscus.
Meniscus: The curved shape of the surface of a liquid in a narrow tube, which is caused by capillarity.
Surface tension: An effect that makes a liquid appear to have a surface "skin."

TRY THIS

Scatter Powder

The surface of water behaves as if it has a skin over it. The skin is strong enough to support bugs and other water insects that can walk on water. But what happens to such floating objects if you stretch the skin?

What to do

Take a large shallow bowl, make sure it is perfectly clean, and fill it nearly to the top with water. When the surface is still and smooth, sprinkle a little talcum powder on it. Put a small smear of liquid detergent on the end of your finger, and gently touch the water near the center of the bowl. The talcum powder will immediately rush away from your finger toward the edges of the bowl.

The detergent breaks the water's surface tension so that it can no longer support the particles of powder. The "skin" on the unaffected water near the edges of the dish pulls away, carrying the powder with it, although it looks as if your finger is pushing it away. Thoroughly wash away all the detergent from the bowl and dry it carefully if you want to repeat the experiment.

Touch the surface of the water with a drop of detergent and watch the powder scatter.

SOLIDS AND CRYSTALS

The expression "solid as a rock" is used to describe anything that is very solid. Dams are made from concrete, which is like artificial rock. It is solid enough to hold back the many millions of tons of water in the lake behind it.

The atoms and molecules that make up solids are hardly able to move at all. They are held together in a regular arrangement. This is responsible for the regular shapes of crystals.

Most solids are hard, rigid, and strong, and many have high melting points. All these properties reflect the fact that their atoms are held together by strong interatomic forces. But some solids consist of large molecules that are held together by weak intermolecular forces. These tend to be soft and have low melting points. Waxes and many polymers (plastics) are typical examples of such solids.

Crystals are solids in which the component atoms or molecules are arranged in a regular, repeating pattern called a lattice. When a crystalline solid is heated, its particles hold their positions in the lattice until the melting point is reached and the solid suddenly melts. A solid that lacks this kind of regular internal arrangement of its particles is termed noncrystalline, or amorphous. Glass and plastics are amorphous solids. When they are heated, they soften gradually over a wide range of temperatures and they have no definite, universally predictable melting point.

The exact nature of the particles in a crystal varies among different materials. In most metals and some other solid elements such as sulfur, or carbon in the form of diamond, the crystals are made up of atoms. In substances such as sugar, the crystal components are molecules. But in the vast majority of crystalline substances the particles are ions. Nearly all salts and minerals—and therefore rocks—are basically ionic solids.

Metals Are Crystals

When we see the word "crystal," we tend to think of something clear and angular, like a sparkling diamond. Most metals are also made up of crystals. This fact, and the type of bonding between the particles within the crystals, explain most of the properties of metals. A typical metal can be cut or polished to produce a shiny surface. Indeed, until people learned how to put a thin layer of silver on a sheet of glass, mirrors were often made from polished metal. Many metals can be drawn out so that the piece gets thinner and thinner to form wire. This property is called ductility, and we say that the metal is ductile. Many metals can also be hammered into thin sheets. Gold, for example, can be beaten into gold leaf so thin that light passes through it. This property is called malleability, and we say that the metal is malleable.

Some metals, such as copper and gold, are both ductile and malleable. They have these properties because, when they are stretched or hammered, their atoms slip past one another so that the solid can take up its

Diamond is among the most valuable of crystals. It is also the hardest solid known. These diamonds have been cut and polished for use in jewelry.

CRYSTAL SYSTEMS

There are seven crystal systems (see page 15), illustrated here by seven common minerals. The external shape is called the crystal habit. All crystals of the same substance always have the same angles between their faces. But the habit may vary because different faces may grow at different speeds.

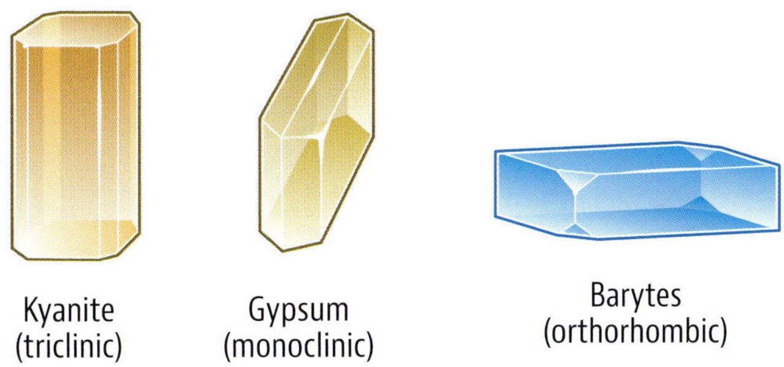

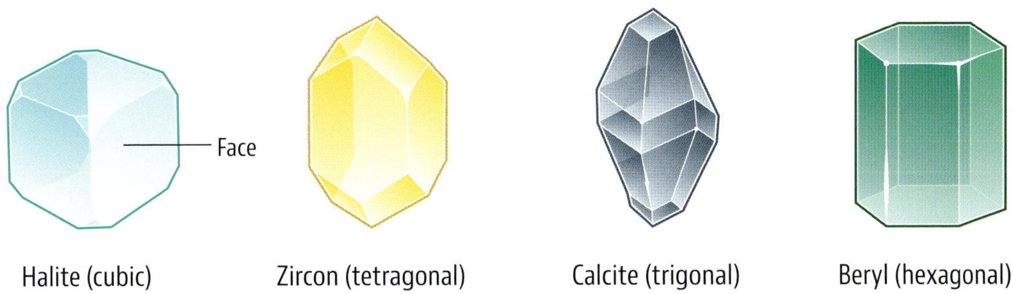

CRYSTAL AXES

The different crystal systems are defined in terms of imaginary axes drawn inside them. Six of the systems have three axes. The simplest is the cubic system, which has three equal axes at right angles. The hexagonal system has four axes.

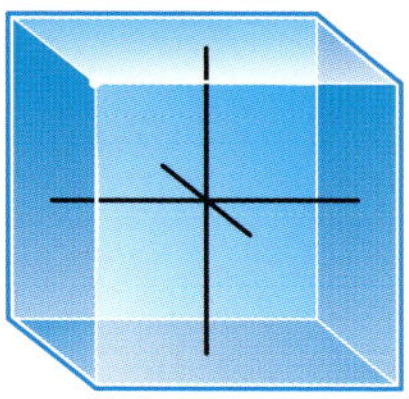

Cubic

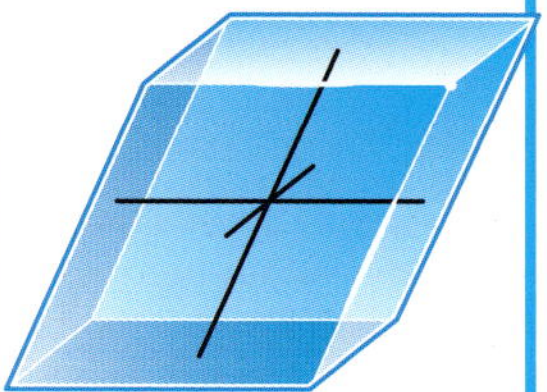

Trigonal

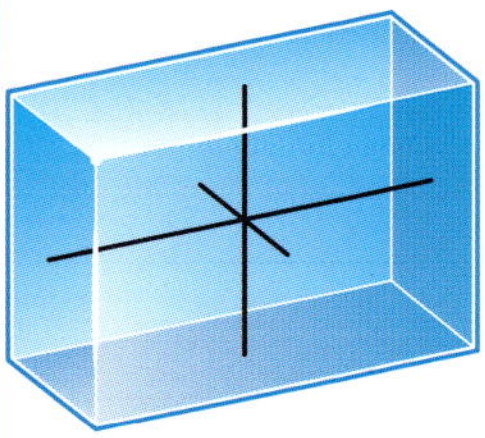

Triclinic

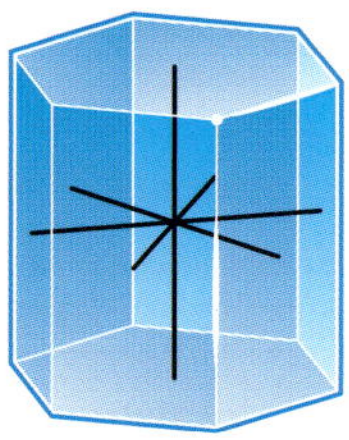

Hexagonal

UNIT CELLS OF CRYSTALS

The basic arrangement of atoms or ions in a crystal is called a unit cell. It determines the outward shape of the crystal. These examples are common among metals and other crystalline substances.

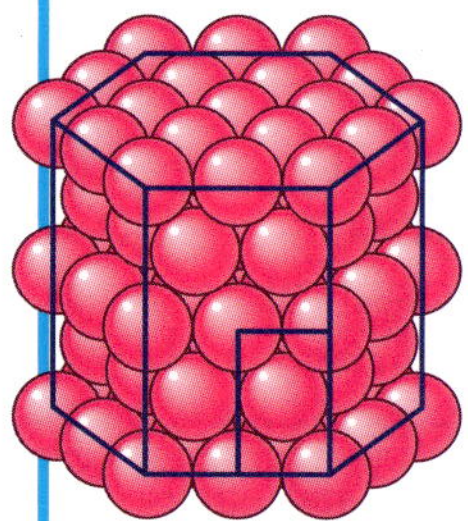

Hexagonal close-packed

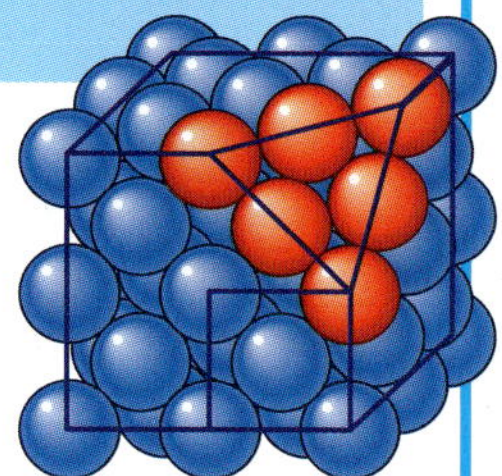

Face-centered cubic

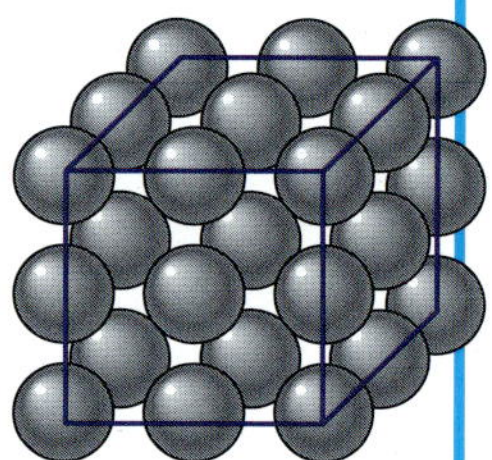

Body-centered cubic

new shape. The kind of bonding in metals must therefore be different from that in other crystalline solids. In a metal, the outermost electrons of the atoms easily become detached. This creates positive ions, which are surrounded by a "sea" of electrons. These electrons are free to move, and this accounts for the ability of a metal to conduct electricity. When a battery, for example, is connected across the ends of a piece of metal wire, the electrons flow along the wire and produce an electric current. The free electrons also make metals good conductors of heat.

Bonding in Solids

We have now identified four different kinds of bonds that hold together the particles in solids. In substances such as diamond, the atoms share electrons and are held together by covalent bonds (see the diagrams on page 4). In substances such as salt, the particles take the form of positive and negative ions held together by electrical attraction called ionic bonds. In a wax, the particles are molecules that are held together by intermolecular forces. And in a metal, the atoms are held together by metallic bonds involving free electrons. However, despite their differences, all four different types of solid can form crystals.

Classifying Crystals

To early scientists, crystals presented a bewildering array of different forms. Natural mineral crystals come in all shapes, sizes, and colors. Gradually, scientists realized that there

SCIENCE WORDS

Crystal: A solid that has a regular shape because of the ordered way in which its atoms or molecules are arranged.
Ductile: Describing a metal that can easily be drawn out to form wire.
Malleable: Describing a metal that can easily be beaten out to form a thin sheet.

are seven basic crystal forms, and all crystals belong to one of them. But natural crystals are seldom perfect and may become distorted while they slowly form in the ground, or they may be affected by the presence of impurities. Throughout the ages, people have valued gemstones such as diamonds, emeralds, and rubies for their beauty and rarity. Even today, collectors eagerly seek perfect specimens of ordinary mineral crystals.

The seven crystal forms are called systems (see page 13). Their names come from their geometrical shape. The simplest is the cube, which is represented in the cubic system. Common salt is found naturally as the mineral rock salt known to mineralogists as halite. It crystallizes as cubes. If you have a strong magnifying glass, you can use it to see the cubic shape of crystals of table salt.

Internal Structure

The particles within a crystal are arranged in a regular pattern called the crystal lattice. Common salt is composed of ions, and in the common salt crystal the ions are arranged at the corners of cubes. Salt crystals are also cubes, so in this case the shape of the crystal reflects the arrangement of ions in the crystal lattice. This is also true for many other crystal shapes. If you imagine eight ions at the corners of a cube, you can see that there is a space in the middle for another ion. This arrangement is the crystal lattice known as body-centered cubic. In another arrangement, there is an extra ion at the center of each face of the cube. This pattern is called face-centered cubic. In yet another arrangement, there is a circle of six ions with a seventh ion in the middle. Picture arranging seven oranges in this fashion. Now imagine adding three more oranges on top of this layer. You could then add another layer of six, and so on, to build up what is called a hexagonal close-packed arrangement.

Scientists look at the structure of crystals to help understand the history of life on earth. They often use x-ray crystallography to study the molecules of carbon. When x-rays pass through a crystal, the atoms or ions scatter the rays and the pattern can be recorded on a piece of photographic film. From these patterns, scientists can work out the crystal structure. In 1951, Rosalind Franklin used this technique to help work out the structure of DNA.

TRY THIS

Grow Your Own Crystals

Mineral crystals form underground when molten material or concentrated solutions gradually cool. The slower they cool, the larger the crystals grow.

What to do

Take a crystalline substance—sugar, salt, or alum work well. Dissolve it in a jar of warm water until no more will dissolve. Tie or fix one crystal to the end of a piece of thin thread and suspend it in the solution (tie the thread to a pencil across the top of the jar). Set it aside for several days. Gradually, the crystal will become larger as it draws more substance out of the solution. The growing crystal keeps the same shape as the original crystal when the new particles take their places in the crystal lattice.

DENSITY AND FLOATING

Which is heavier: a piece of wood or a piece of steel? The answer depends on their sizes. A piece of wood the size of a table is heavier than a piece of steel the size of a pin. But a ton of wood weighs the same as a ton of steel.

A piece of steel may or may not be heavier than a piece of wood, depending on their sizes. But steel is always denser than wood. The density of a substance is its mass divided by its volume. Steel has a density of about 480 lb per cu ft (7,800 kg per cu m), while the density of wood is only about 44 lb per cu ft (700 kg per cu m), depending on the type of wood. On average steel is about 11 times as dense as hard wood.

Water has a density of 62 lb per cu ft (1,000 kg per cu m), higher than wood but less than steel. That is why a piece of wood will float on water while a lump of steel will sink. This difference in densities is also why a balloon filled with helium gas (density 0.01 lb per cu ft/0.18 kg per cu m) floats in air (density 0.08 lb per cu ft/1.3 kg per cu m). For something to float, it must be less dense than the fluid—liquid or gas—it is floating in. Ice is less dense than water, which is why ice cubes float in a drink and icebergs float in the ocean. But ice is only a little less dense than

Steel-hulled ships like this container carrier float because they are hollow. A ship displaces its own weight of water. It rides higher in the water when it has no cargo on board and sinks deeper when it is fully loaded because it weighs more.

ICEBERGS FLOAT

Icebergs float because ice is less dense than seawater. The density of ice is about nine-tenths the density of water. For this reason, only 10 percent of the floating ice is above the water. The remaining 90 percent remains submerged beneath the surface.

seawater, which is why the greater mass of an iceberg stays below the surface, often jutting out underwater and endangering any ships that get too close to the iceberg.

Sometimes scientists use relative density instead of density. It is equal to the density of a substance divided by the density of water. It is a pure number with no units. Gold has a density of 1,190 lb per cu ft (19,000 kg per cu m) and a relative density of 19 (equal to the density of gold divided by the density of water).

Density and Strength

Steel is a strong metal alloy (combination of metals). Aluminum is not quite as strong, but it is less dense (170 lb per cu ft, or about 2,700 kg per cu m). For this reason, aluminum alloys are used in space rockets and airplanes. A steel airplane would be three times as heavy as an aluminum one, and its engines would have to be three times as powerful to lift the airplane off the ground.

But some materials that have a low density are also strong. That is particularly true of composite materials such as fiberglass and carbon fiber. Composite materials are made

SCIENCE WORDS

Density: For any substance its mass divided by its volume.
Upthrust: The apparent loss in weight of a floating object, equal to the buoyant force keeping it afloat.

by combining two or more different materials to make a new material that is better (usually stronger) than the ones it is made from. Fiberglass has thin, hairlike strands of glass set in a plastic resin. It is slightly stronger than steel, but only one-fourth its density. It is used for making the hulls for small boats, for fishing rods, and for the poles used by pole-vaulters. Carbon fiber is a similar material but is even stronger.

Floating and Displacement

One of the oldest stories in physics is about a Greek mathematician and scientist called Archimedes, who lived more than 2,200 years ago. According to the story, he decided to take a bath and climbed into a tub that was full to the brim with water. As soon as he got into the bath, the water overflowed onto the floor. But Archimedes (c. 287–212 BCE) was so pleased with this discovery that he ran out into the street crying "Eureka!" (Greek for "I have found it!"). What Archimedes had discovered was that an object immersed in water displaces its own volume of water. He also realized that the force keeping an object afloat—the so-called upthrust, or buoyant force—is equal to the weight of the water (or whatever fluid the object is floating in) that it displaces. This is known as Archimedes' Principle. Any object that weighs more than the buoyant force will sink, but an object that weighs less than the buoyant force will float.

Another story about Archimedes shows how he put his principle to good use. Archimedes lived in the city of Syracuse on Sicily. One day the king of Syracuse asked for his help. The ruler had been given a golden crown, and he asked Archimedes if he could find out whether it was made from pure gold without melting it down and destroying it. Archimedes first weighed the crown and then immersed it in a bowl of water and measured the volume of water displaced. This volume of water equaled the volume of the crown, so Archimedes divided the weight of the crown by its volume, which gave him the density of the metal. He knew the density of pure gold, so he knew if the crown was made from pure gold. Sadly, history does not record whether the king's crown was made from pure gold or not!

ARCHIMEDES' PRINCIPLE

When a weight is suspended in water, it appears to weigh less by an amount equal to the weight of water that it displaces.

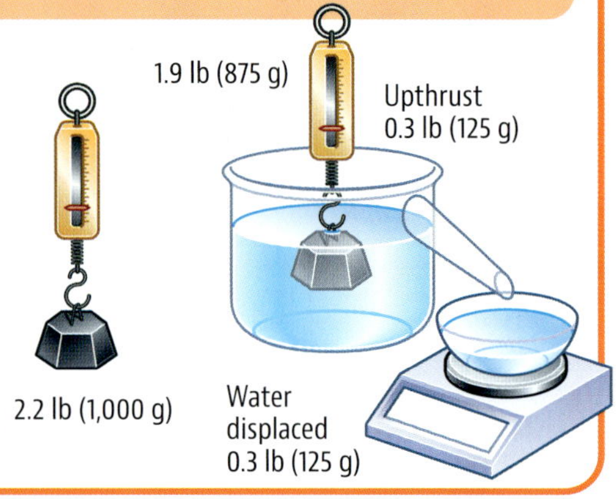

ARCHIMEDES

Archimedes was a mathematician, engineer, and scientist from ancient Greece. He lived in Syracuse on the island of Sicily. In mathematics, he found new methods for measuring areas and volumes, and developed ways of dealing with very large numbers. He invented a pump that used a rotating, angled screw for lifting water, and made pulley blocks for lifting heavy weights. However, he is best known for discovering Archimedes' Principle, which states that the upward force on an object immersed in a liquid is equal to the weight of the liquid displaced. A Roman army attacked Syracuse in 215 BCE, and Archimedes was killed three years later by a Roman soldier.

When the air inside a hot-air balloon is heated, it becomes less dense than the air outside it, and the balloon rises into the sky.

Ships and Submarines

Although steel is much denser than water, we can now begin to understand how a steel ship floats. A solid lump of steel always sinks in water, but the hull of a steel boat is hollow. In water, the hollow hull displaces an equal volume of water. The weight of water displaced equals the weight of the ship's hull, and so the ship floats. The weight of a ship is usually expressed as so many tons displacement, which is the weight of water it displaces. Some modern supertankers have displacements in excess of 500,000 tons. In a similar way, airships displace their own volume when they "float" in the air. The air in a hot-air balloon is less dense than the surrounding air that it displaces as it floats.

A submarine also has a hollow hull, but it is surrounded by tanks called ballast tanks. On the surface, the tanks are empty. The submarine floats. To make it submerge quickly, water is allowed into the tanks. The vessel loses buoyancy and slowly sinks. To help it surface again, compressed air forces the water out of the tanks. The vessel becomes buoyant again and rises. The principle is the same for a small submersible and for a large nuclear submarine.

SUBMARINES

A submarine sinks or rises vertically in the water as its buoyancy changes. Letting water into the ballast tanks or forcing the water out with compressed air alters the buoyancy.

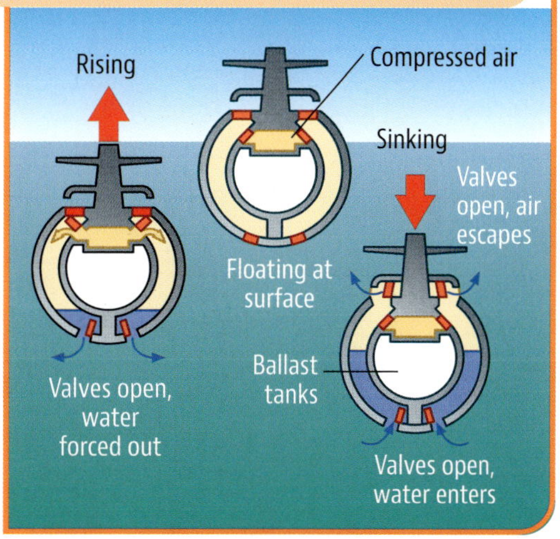

HEAT AS ENERGY

Heat is a form of energy. Hot things possess heat energy, which they "store" in their vibrating atoms or molecules. The more these particles vibrate, the hotter the material gets. Particles in cold materials do not vibrate as much.

Energy comes in many different forms. For example, light and electricity are both forms of energy. Objects in motion possess kinetic energy, and even stationary objects have potential energy. In addition, every object contains some internal energy that lies in the small vibrations of the atoms or molecules it is made from. It is this energy that we call heat. More technically, heat represents the transfer of energy from a hot object to a cooler one because of their difference in temperature.

A red-hot nail is much hotter than a bucket of warm water. Yet the water contains more heat energy than the nail. The level of "hotness" is an object's temperature, and it depends on how vigorously the object's component particles—its atoms or molecules—are vibrating. The bucket of warm water contains many more particles than the nail does, and as a result it "stores" more heat, even though its temperature is much lower. Adding heat to an object will increase its temperature, while removing heat from it will have the opposite effect, that is lowering its temperature.

SCIENCE WORDS

calorie: (cal) A unit of heat equal to the amount of heat needed to raise the temperature of 0.035 oz (1 g) of water through 33.8°F (1°C).
Calorie: (with a capital C) 1,000 calories, the same as a kilocalorie.
Joule: (J) Unit of energy, equal to a force of 1 newton through a distance of 3.2 ft (1 m).

Units of Heat

Because heat is a form of energy, its correct scientific unit is the joule. Technically, 1 joule is the amount of work done when a force of 1 newton acts through a distance of 3.2 ft (1 m). All forms of energy—and work—can be measured in joules.

An older unit called the calorie (cal) is sometimes used to measure heat. One calorie is the heat needed to raise the temperature of 0.035 oz (1 g) of water through 33.8°F (1°C). The calorie is a small unit, and often the kilocalorie is used in practice. One kilocalorie

A grass fire quickly gets out of control as it is spread by the wind. It is an example of the potential destructive power of flames.

(kcal) equals 1,000 calories. To add to the confusion, dieticians who deal with energy values in foods call this unit the Calorie (with a capital C); 1 Calorie equals 1,000 calories. Often the capital C is omitted on food labels. So a 200-calorie piece of candy actually has an energy content of 200 kcal—more than enough to boil just over 2 quarts (2 kg) of water.

Sometimes it is necessary to convert calories into joules (or vice versa). You need to know that 1 calorie equals about 4.2 joules (4.1868 joules, to be precise).

TRY THIS

Warmer Is Wetter

Warm water is "thinner" than cold water, and as a result it flows more easily. Here is a simple demonstration.

What to do

Wash out two paper cups thoroughly if they have been used. Turn the cups upside down, and make a small hole in the center of the base of each cup with a pin. Carefully fill one cup about three-fourths full with very hot water from the hot-water faucet. Put the paper cup on top of a flat-bottomed glass. Put the same amount of cold water in the other paper cup and add an ice cube. Put this cup on top of another glass. Now watch what happens.

The hot water will drip through the hole faster than the cold water. This is because it is "thinner"—in scientific terms it is less viscous (more viscous liquid is thick, like syrup). The molecules in the hot water move much more quickly than those in the cold water. For this reason, they slip past one another and through the hole more easily.

Hot water drips through a small hole quicker than cold water does.

PRODUCING HEAT

Heat can be produced in many ways—by burning things, by using electricity, even by rubbing your hands together. You can also feel heat in the Sun's rays, while the Earth's core is a source of heat, which can be tapped to generate geothermal power. Nuclear reactors also produce heat.

SCIENCE WORDS

Friction: A force that prevents or slows the movement of one surface against another surface. It results in the production of heat.
Geothermal energy: A form of heat energy that comes from deep underground, such as geysers, hot springs, and volcanoes.
Heat: The internal energy of an object that results from the vibrations of its particles (atoms or molecules).

Because heat is a type of energy, it is generally produced by converting other forms of energy. For example, when a piece of coal burns, the chemical energy in the coal is released as heat in a chemical reaction called combustion (burning). Chemical energy is converted to heat when any fuel burns, and it is often put to use in engines such as the diesel engine in a truck.

Another source of heat for homes and industry is electricity. Electric heaters rely on the fact that when an electric current flows through a piece of wire, the wire gets hot. Special wire is used so it makes heat without melting or degrading in the air.

Mechanical Energy Into Heat

The conversion of mechanical energy into heat was observed in the 1790s by Count Rumford (1753–1814). He correctly assumed that the mechanical energy of a machine boring a cannon barrel was converted to heat because of friction between the drill and the metal of the barrel. He even submerged the barrel and the drill in a tank of water, and after a while the water in the tank boiled because of the heat produced.

Heat caused by friction creates major problems in any kind of machinery that has moving parts. Engineers design special

HEAT PRODUCTION

Six sources of heat: combustion (burning), electricity, the Sun, friction, geothermal, and nuclear.

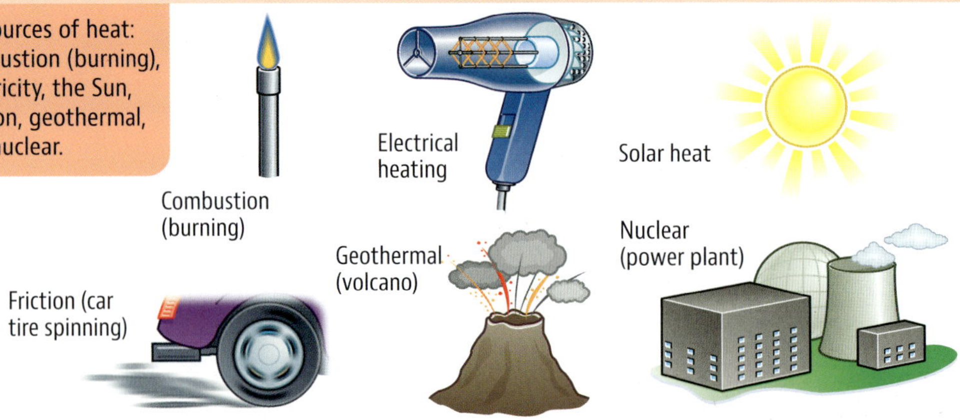

The back tires of a drag racer spin at the beginning of a run. Friction makes the surface of the tires hot so that they burn and soften as the car streaks away.

bearings for rotating shafts to minimize friction, which can also be reduced by using a lubricant such as oil. Also, many drilling and milling machines have a watery lubricating liquid that flows over the workpiece and carries away the heat produced by the friction between the parts.

Other Sources of Heat

Various natural objects are hot and can give up some of their heat. Such objects include the Sun, (with heat you can feel on your skin) and the Earth. The Earth's core, at a temperature of about 7,200°F (4,000°C), is surrounded by a 1,850-mile (3,000-km) thick mantle of molten rock, topped by a rocky crust. By drilling down into the crust, which is up to 25 miles (40 km) thick, engineers can tap the Earth's internal heat. Hot springs and volcanoes also help to bring this geothermal energy to the surface.

COUNT RUMFORD

Count Rumford was born Benjamin Thompson in 1753 in Woburn, Massachusetts. He was the son of a farmer, and became an international politician and physicist. He spied for the British during the American Revolutionary War (1775–1783), and in 1785 went to Germany, where he became a count.

Watching machines boring cannon barrels at the Munich arsenal, Rumford noticed how hot the barrels became. He realized that heat is a form of energy that is produced by friction. After his return to England in 1798, Count Rumford invented several devices, including a kitchen stove, an oil lamp, and a photometer, which he used for measuring the brightness of light.

HOTNESS AND TEMPERATURE

We all know that something with a high temperature is hot. Temperature is the degree of the hotness of something, which is a measure of how vigorously its atoms or molecules are vibrating. Scientists express temperatures on a temperature scale.

TEMPERATURE SCALES

The chief temperature scales are the Fahrenheit scale, used for weather forecasts and everyday purposes; the Celsius scale (formerly called centigrade), used in science and for everyday measurements in most European countries; and the Kelvin scale, which is based on absolute zero. A Kelvin degree is the same size as a Celsius degree.

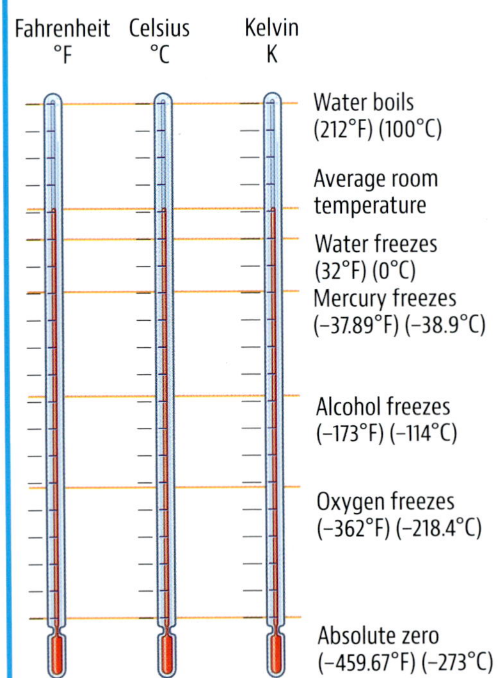

Blast furnaces producing molten steel can reach temperatures of 3,600°F to 4,200°F (2,000 to 2,300°C).

We know that if we want to make something hotter, we have to supply it with heat. It might be reasonable to assume that you could make something hotter by taking a warm object and putting it in contact with a hot object thinking that the heat from the warm object will travel to the hot one—making it even hotter. However, this goes against the laws of physics. One of these laws states that heat will not flow of its own accord from a cooler object to a warmer one. It will only flow from a warm or hot object to a cooler one. To measure an object's

temperature we use some form of thermometer. The thermometer has to be graduated in degrees of hotness. These graduations together form a temperature scale, and over the centuries various scales have been devised. Most have to have at least two fixed points, such as the temperatures at which water freezes and boils.

Two common temperature scales are named after their inventors. The German physicist Gabriel Fahrenheit (1686–1736), who made the first alcohol-containing thermometer in 1709, invented the Fahrenheit scale. On this scale, the freezing point of water is 32°F (0°C), and the boiling point is 212°F (100°C). The Celsius scale was devised by the Swedish astronomer Anders Celsius (1701–1744) in 1742. It has the freezing point of water as 32°F (0°C) and the boiling point as 212°F (100°C). Since there are 100 degrees in this range, it was originally called the centigrade scale ("centi" meaning a hundred).

Thermodynamic temperature is based on the motion of atoms and molecules. The lowest temperature possible is called absolute zero, which equals –459.4°F (–273°C). At this temperature atoms and molecules stop moving altogether, so nothing can be made any colder. Scientists measure thermodynamic temperature on the Kelvin scale, on which absolute zero is 0 K (notice there is no degree sign). On this scale, the freezing point of water is 273 K.

The Kelvin degree is the same size as a Celsius degree. Temperature differences—as opposed to fixed temperatures—are often expressed in Kelvin rather than in Fahrenheit or Celsius. The Kelvin scale is named for the Scottish physicist William Thomson (1824–1907), who became Lord Kelvin in 1892 when he was honored for his services to science.

SCIENCE WORDS

Absolute temperature scale: The temperature scale that begins at absolute zero. It is also called the Kelvin temperature scale.

Celsius scale: A temperature scale that has 100 degrees between the freezing point of water (0°C) and the boiling point of water 100°C. This scale used to be called the centigrade scale.

Centigrade scale: An old name for the Celsius scale.

Fahrenheit scale: A temperature scale that has 180 degrees between the freezing point of water (32°F) and the boiling point of water (212°F).

LATENT HEAT

It takes a certain amount of heat to raise the temperature of water to its boiling point. If heat continues to be applied to the boiling water, it converts the water to steam. This extra heat is called latent heat.

Whenever a solid melts to form a liquid, or a liquid boils to form a gas or vapor, we say that a change of state has occurred. Similar changes happen when a gas or vapor condenses to form a liquid, or when a liquid freezes to form a solid.

Every change of state involves the absorption or release of heat. It takes heat to melt ice—not just to raise its temperature but also to change it from a solid to a liquid. This heat is called latent heat, and for ice it takes 80 kcal to melt 2.2 lb (1 kg). Its temperature remains at freezing point, 32°F (0°C). Scientists describe this latent heat of fusion of ice as being 80 kcal/kg. Not surprisingly, the latent heat of freezing of water is exactly the same. To convert it into ice you have to remove 80 kcal from 2.2 lb (1 kg) of water. The latent heat of vaporization also applies to the change from liquid to gas or vapor.

For water, it takes 540 kcal to change 2.2 lb (1 kg) of liquid water at 212°F (100°C) into 2.2 lb (1 kg) of steam at the same temperature. The latent heat of condensation, which is the amount of heat that must be removed to convert 2.2 lb (1 kg) of steam at 212°F (100°C) to 2.2 lb (1 kg) of liquid water at 212°F (100°C), is also 540 kcal. When a liquid evaporates, it takes heat from its surroundings. That is why sweating makes you cooler. The latent heat of vaporization of the watery sweat is removed from your skin.

Solid to Vapor

Some solids are unusual because when they are heated they turn directly into a vapor without melting first. This process is given the name sublimation, and the solid is said to sublime. It occurs because the boiling point of the solid is less than its melting point at atmospheric pressure. Examples of substances

When a kettle boils, supplying more heat turns the boiling water into steam. This property is called latent heat of vaporization.

that sublime are solid carbon dioxide ("dry ice") and iodine. At any temperature above –109°F (–78.3°C), solid carbon dioxide changes directly into a gas. Carbon dioxide and iodine sublime at ordinary atmospheric pressure. Other solids can be made to sublime by greatly reducing the pressure. This is the principle of the freeze-drying process for foods. The food is frozen at low pressure so that the ice sublimes out of it, leaving behind a dehydrated product. Conversely, at high pressures a substance that sublimes at atmospheric pressure can be made to melt. Liquid carbon dioxide can be produced at high pressures.

SCIENCE WORDS

Latent heat: The heat taken in or given out when a substance undergoes a change of state. The latent heat of fusion is the heat needed to change a solid at its melting point into a liquid (and is the same as the latent heat of freezing). The latent heat of vaporization is the heat needed to change a liquid at its boiling point into a vapor (and is the same as the latent heat of condensation).

Vaporization: The change from liquid to gas or vapor on heating.

HIDDEN HEAT

The word "latent" means "hidden." This diagram shows where the hidden heat goes. First, it takes 80 kcal of heat to melt 2.2lb (1 kg) of ice supplied with heat at the rate of 10 kcal per minute. When the water has been heated to its boiling point, it takes another 540 kcal to turn 2.2lb (1 kg) of water into 2.2lb (1 kg) of steam.

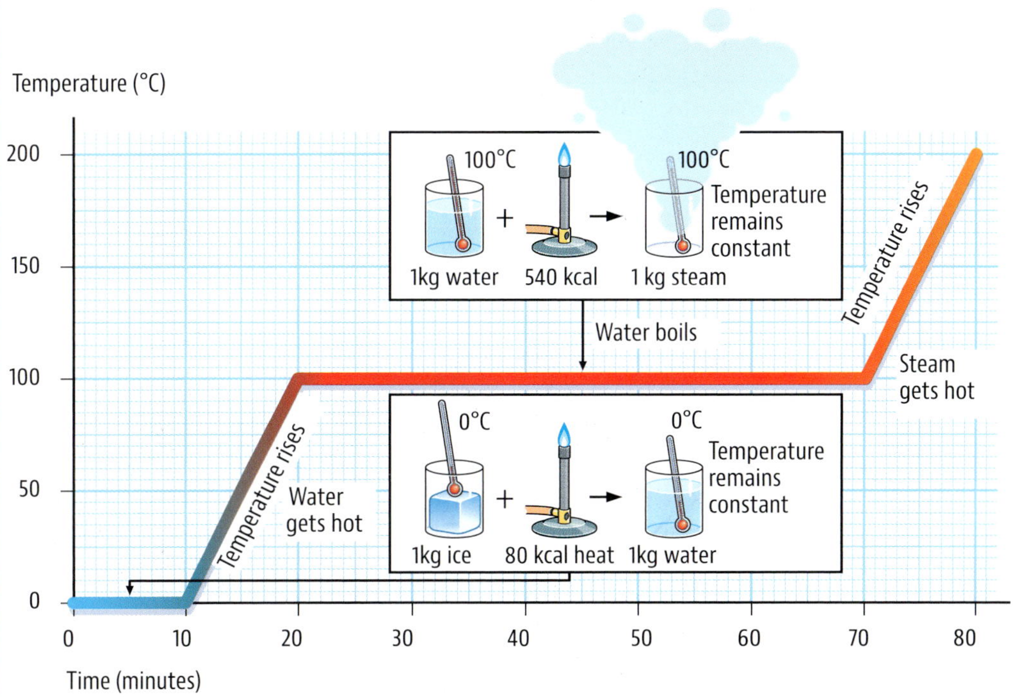

EXPANSION OF FLUIDS

To a physicist, a fluid is any gas or liquid, such as air, steam, or water. These two states of matter are grouped together because they share a common property: they can both flow—hence the name. Fluids also share another property: they expand when they are heated. This property is used in many kinds of thermometers.

EXPANSION OF MERCURY

A teaspoonful of mercury equals about 0.2 oz (5 ml), which is the volume of mercury in a large laboratory thermometer. This illustration shows how much the volume of mercury expands when it is heated to 122°F (50°C) and 212°F (100°C).

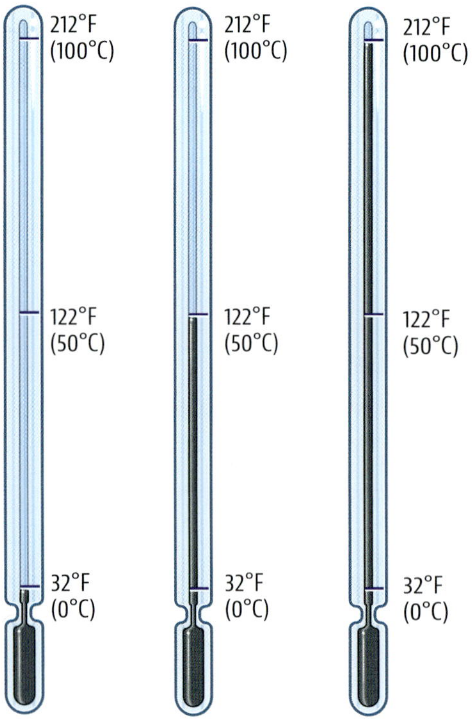

Liquids or gases have no definite shape—they take on the form of the vessel containing them. Therefore only volume expansion can really be measured.

For a liquid, expansion is measured as its coefficient of volume expansion, which is the proportion by which the original volume expands when the temperature is increased by 1 K at a time. These coefficients are very small, but still up to ten times larger than those for solids, and they vary slightly depending on the temperature at which they are measured. For example, water has a coefficient of volume expansion of 0.0003 per K, while that of mercury is even smaller at 0.000018 per K.

Measuring Liquid Expansion

It is relatively straightforward to define the coefficient of volume expansion of a liquid, but it is quite another matter to try to measure it. That is because the liquid has

to be held in a container of some sort, and attempts to measure the expansion of the liquid on heating has to take into account the expansion of the container as well.

This difficulty can be demonstrated by making a mark on the side of a glass beaker just level with the surface of a liquid. When the beaker is heated from below, the level of the liquid initially falls as the beaker expands. When the heat reaches the liquid, it expands and its level rises above the mark. As a result, it is only possible to measure the apparent expansion of the liquid. For this reason, physicists sometimes distinguish between a liquid's absolute, or it true, coefficient of expansion and its apparent coefficient of expansion. There are ways of getting around this difficulty if we know the coefficient of volume expansion of the material of the beaker, in this example, glass.

The expansion of hot gases is what propels a shell out of a gun, demonstrated here by a tank firing its cannon.

TRY THIS

Climbing Water

In this project, you will make a liquid expand and climb vertically up a tube.

What to do

Remove the top from a plastic bottle. Place the bottle top, flat part upward, on a stone or concrete surface, or on a brick. Using a hammer and nail, carefully make a hole in the center of the top. Push a drinking straw through the hole. If the hole is not big enough, gently enlarge it by turning the blades of a pair of closed scissors in the hole.

Half-fill the bottle with very cold water containing a little food coloring. Push the straw down through the hole in the bottle top until it nearly touches the bottom of the bottle when the top is screwed back on. Pack modeling clay around the straw to seal the hole in the bottle top. Stand the bottle in a bucket or bowl. Use a jug to fill the bucket with hot water to the same level as the colored water in the bottle and watch what happens.

As the colored water in the bottle is heated by the hot water in the bucket, it expands. Because the top end of the drinking straw is open, the colored water moves up the straw. It may even reach the top and form a miniature colored fountain. Movementof a warmed liquid along a tube in this way is how an ordinary thermometer works.

The colored water rises up the tube and may even spurt out as a miniature fountain.

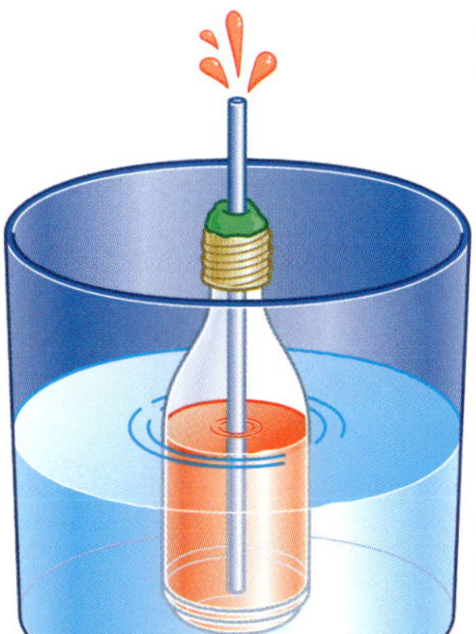

Exceptional Water

In the same way that ice is an unusual solid because it expands as it cools, water is also exceptional in the way that it behaves. When cold water is warmed from a temperature of 32°F (0°C) to 39.2°F (4°C), it takes up less volume because it contracts. Above 39.2°F (4°C) it begins to expand, behaving in the same way as most other liquids.

Because density equals mass divided by volume, and warming does not change the mass of water, water must be at its densest at about 39.2°F (4°C). This effect has some important consequences for fish. Since water near the surface of a lake cools toward the freezing point in winter, its density increases, and it sinks to the bottom. Eventually, the surface water reaches a temperature of 39.2°F (4°C); but then, as it cools further, the colder water remains at the top—because it is less dense than the water lower down.

SCIENCE WORDS

Coefficient of expansion: A measure of how much a substance expands when it is heated.

Coefficient of volume expansion: A measure of how much the volume of a substance (solid, liquid, or gas) increases when it is heated.

Even when the surface freezes and forms ice, the water at the bottom of the lake remains liquid at a temperature of 39.2°F (4°C), providing a safe home for fish.

Heating increases the volume of a gas as long as its pressure is kept the same. Like a liquid, a gas has a coefficient of volume expansion defined as the increase in volume of 1 cu cm of a gas at 32°F (0°C) for a rise in temperature of 1 K (at constant pressure). It turns out to have the same value for all gases

CHARLES'S LAW

The volume of a gas increases by 1/273 of its volume at 32°F (0°C) for every degree rise in temperature. Therefore a balloon with a volume of 10 cu m at 32°F (0°C) increases in volume to 15 cu m at 278.6°F (137°C) and doubles in volume to 20 cu m at 524°F (273°C).

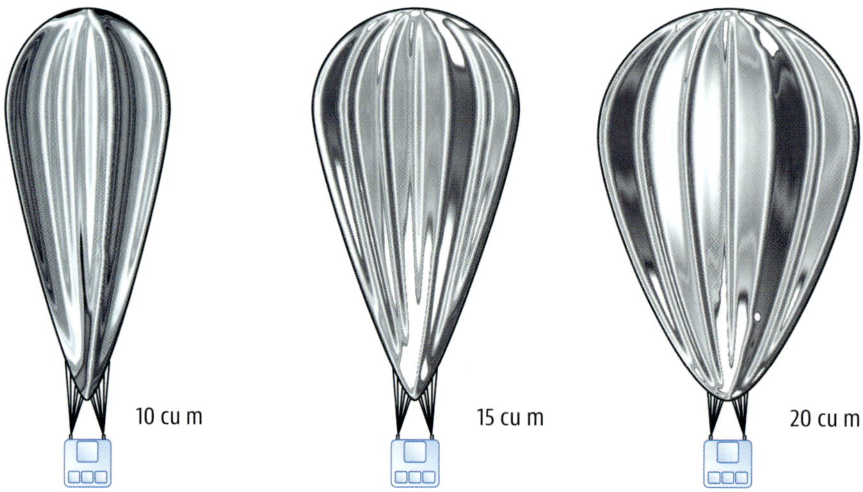

and is equal to 0.00366 per degree K. This is equal to the fraction 1/273, as predicted by Charles's Law. The law states that at constant pressure the volume of a given mass of gas increases by 1/273 of its volume at 32°F (0°C) for every degree Kelvin rise in temperature. As a result, the volume of a gas at a constant pressure is proportional to its absolute temperature: that is, its temperature on the Kelvin temperature scale. Degrees Kelvin equals degrees Celsius plus 273.

Gases have another coefficient of expansion, which is measured in terms of changes in gas pressure when the volume is kept constant. It is called the pressure coefficient and turns out to be equal to 0.00366 per K (or 1/273 per K), the same as the volume coefficient.

This fact should come as no surprise because both coefficients represent the effect of heating on the molecules. At constant pressure, the molecules acquire more energy and move farther apart, so increasing the volume. At constant volume, the more energetic molecules collide with the walls of the container more frequently and thus increase the pressure.

Putting Gases to Work

The pressure created by heating a gas can be made to do work. For example, gas pressure is used in steam engines, internal combustion engines (gasoline and diesel), gas turbines, and pneumatic tools. Hot expanding gases are also the propellants in rockets and guns.

Balloons also contain gas. In a hot-air balloon, air is heated so that it expands and becomes less dense than the surrounding air, making the balloon rise. Expansion caused by heating also has to be taken into account with gas-filled balloons, as explained in the illustration of Charles's Law, opposite.

TRY THIS

The Dime That Dances

Heat makes solids, liquids, and gases expand (get bigger). This project illustrates a strange effect produced by expanding air.

What to do

Place a dime over the top of a bottle. Dip your finger in some water and drip a little around the edge of the coin to make a seal. Grip the bottle in both hands for about 20 seconds, being careful not to dislodge the dime. The coin will start to move up and down, making a clicking sound. Keep hold of the bottle for a while, then put it down. The dime will go on dancing.

When you grip the bottle, the heat from your hands warms the air inside the bottle. The warmth makes the air expand. The air has to push past the dime to escape from the bottle. The dime goes on dancing until the air pressure inside the bottle is the same as the air pressure outside it.

As the warm air escapes it makes the dime dance on the top of the bottle.

BOILING AND EVAPORATION

When a liquid is heated above a certain temperature, it boils and turns into a gas, usually called a vapor. Steam is the gas formed when water boils. Liquids also turn to vapor when they evaporate, even though they are not boiling.

The molecules in a liquid are free to move around, allowing a liquid to flow and take up the shape of its container. When the liquid is heated, the extra energy makes the molecules move ever faster and move farther apart. Eventually, at a certain temperature called the boiling point, the molecules are so far apart that they become a gas. Usually the gas first forms within the liquid near the base of the container that is being heated. It takes the form of bubbles, which move rapidly upward and make the liquid froth. Boiling is complete when the bubbles reach the surface and burst.

Different liquids boil at different temperatures. The boiling point of water, for example, is 212°F (100°C), while that of ether is only 94°F (34.5°C). That is so low that a drop of ether will boil if placed on the palm of your hand. The liquid metal mercury does not boil until it reaches about 676°F (358°C). That makes it a good liquid to use in thermometers. The metal tungsten, which was once widely used to make electric lamp filaments, has the highest boiling point of all. It boils at an incredible 10,220°F (5660°C), which is as high as the temperature on the surface of the Sun.

Changing Boiling Point

Various things can alter the boiling point of a liquid. A liquid boils when the pressure of the gas inside the bubbles is the same as atmospheric pressure—the pressure of the air at the surface of the liquid. If we change the pressure, the boiling point changes. The air

LEAPING WATER MOLECULES

In a boiling liquid **(a)**, high-energy "hot" molecules leap out of the surface and form bubbles of vapor in the liquid. In evaporation **(b)**, high-energy molecules leave the surface of the liquid even though it is not heated. As a result, the liquid loses heat energy, and it gets cooler **(c)**.

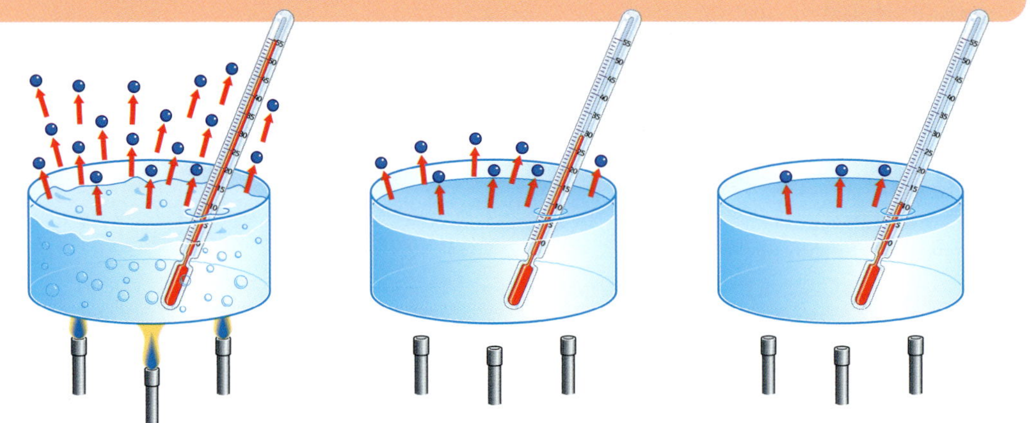

(a) Boiling: molecules driven off **(b)** Evaporation: molecules escape **(c)** Evaporation causes cooling

pressure at the top of a high mountain is much less than it is at sea level. As a result, water boils at a lower temperature at the mountaintop. For example, at an altitude of about 10,000 ft (3,000 m) water boils at 194°F (90°C). Mountaineers find it difficult to make a good cup of coffee at the top of a mountain because the pressure is lower and the water does not get hot enough.

The 18th-century American scientist Benjamin Franklin (1706–1790) did a simple experiment that demonstrated the effect of pressure on boiling water. He heated water in a flask until the water boiled, then sealed the flask with a cork. As the water cooled the boiling stopped. He then turned the flask upside down and poured cold water over it. Some of the steam above the water inside the flask condensed, making the pressure inside the flask less than the atmospheric pressure outside it. As a result, the water inside started to boil again.

Increasing the pressure on a liquid makes it boil at a higher temperature. This principle is used in pressure chambers for sterilizing surgical instruments, and in pressure cookers. If the pressure in the cooker rises to twice the atmospheric pressure, the water boils at a temperature of 248°F (120°C). As a result, food cooks much quicker. At 15 times the atmospheric pressure, water does not boil until the temperature reaches 392°F (200°C).

An erupting hot spring or geyser is an impressive sight. Below the geyser, groundwater flows down a channel toward very hot underground rocks. The heat boils the water, which forms steam. The pressure builds up until the steam explodes back up the channel, along with any water in its way.

The boiling point of a liquid also depends on its purity. An impure liquid boils at a higher temperature than a pure liquid. That is why adding salt to water raises its boiling point: potatoes cook quicker in boiling water with salt added. How much the boiling point rises depends on the concentration of the added substance. Chemists measure the rise in boiling point to find the concentration of a solution.

Evaporation

Heating a liquid until it boils turns it to gas or vapor, but vapor can form without any heat

SCIENCE WORDS

Atmospheric pressure: The pressure of the Earth's atmosphere at any point on the Earth's surface—caused by the weight of the column of air above it. Atmospheric pressure decreases with altitude.
Boiling point: The temperature at which a liquid changes into a gas or vapor.
Condensation: The process by which a gas or vapor changes into a liquid. The liquid formed is also sometimes called condensation.

THE WATER CYCLE

The water cycle describes how the Earth's water travels between the clouds and the oceans and rivers. The oceans always contain about 97 percent of the water, with only 0.014 percent being in lakes and rivers. Clouds account for about 0.001 percent. Most of the remainder is ice locked up as ice in polar ice caps and glaciers.

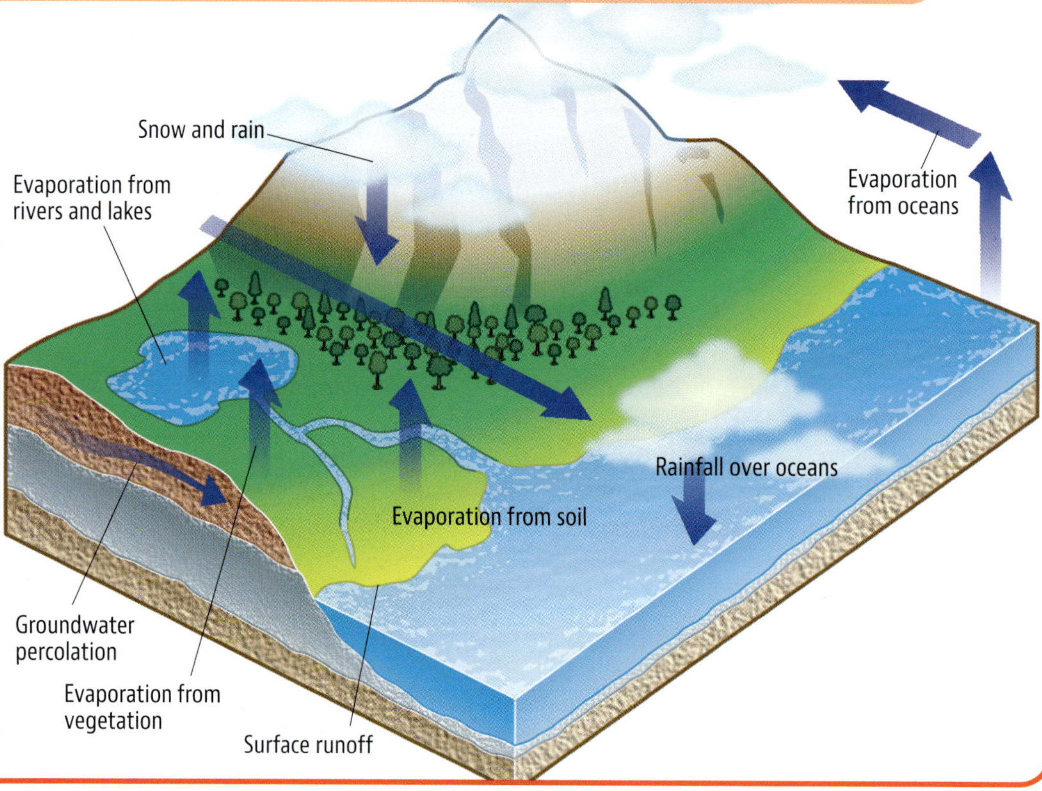

being applied. A puddle that forms after it has rained gradually dries up, but where does the water go? The answer is it turns to vapor. We know that the molecules in a liquid move around. At the surface of the liquid, some of the faster molecules leap right off the surface, become vapor, and never return. This process is called evaporation. It can happen even at low temperatures, but it speeds up as the temperature rises.

The faster molecules leaving the surface during evaporation take heat energy with them. As a result, the liquid that remains gets cooler. That is why laundry hanging out to dry feels cold as the water evaporates from it. Evaporation also accounts for the cooling effect of sweating. When a human or other animal sweats, the liquid perspiration evaporates, taking heat with it.

A simple device for keeping drinks cool works by the same principle (see "Try This," right). It consists of a porous pot, such as an unglazed flowerpot, soaked in water and placed over the drink carton or can of soda. As the water evaporates from the pot, it takes the heat and keeps the drink cool. Some wine bottle coolers also work in the same way.

Water Cycle

Everyone knows that rainwater comes from clouds, but where did the water come from? It all depends on the evaporation of water from rivers, lakes, and oceans.

Water rises into the sky as water vapor, where it cools and condenses into tiny droplets of water. The living processes of plants also give off water vapor, adding to the vapor in the sky. Clouds are blown around by the wind, but eventually the water droplets in the clouds become large enough to fall as rain. If it is cold enough, they fall as snow. Most melted snow and rainwater forms streams and rivers that eventually flow back to the sea. Some water also evaporates from the ground, while other water soaks into the ground and may reemerge in springs. Some of the groundwater is taken up by plants and then passes back into the air from their leaves as water vapor.

The whole process whereby water travels between the clouds and oceans and rivers is called the water cycle.

TRY THIS

Keeping Your Cool

How can you keep a cold drink cool in the summer without an icebox? One way is to apply a scientific method that has been in use for hundreds of years.

What to do

Soak an unglazed clay flowerpot in a bucket of water overnight. The next day, pour some cold water into a bowl, place a soda bottle or can in the bowl and cover it with the upside-down flowerpot. Put a stone over the hole in the base of the pot. The drink will stay cool.

What happens is that water evaporates from the surface of the pot. As it does so, it draws heat from its surroundings. Because the pot is porous, it soaks up water from the bowl like a sponge, keeping the pot wet so that evaporation keeps on going. Taste the drink from time to time to test that it is keeping cool!

Water evaporating from the surface of the flowerpot keeps the trapped air cool.

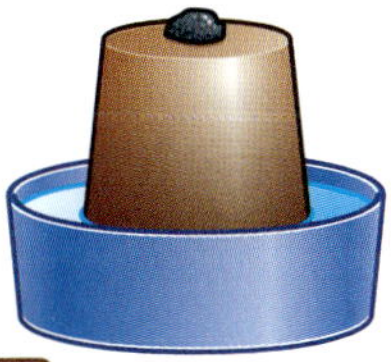

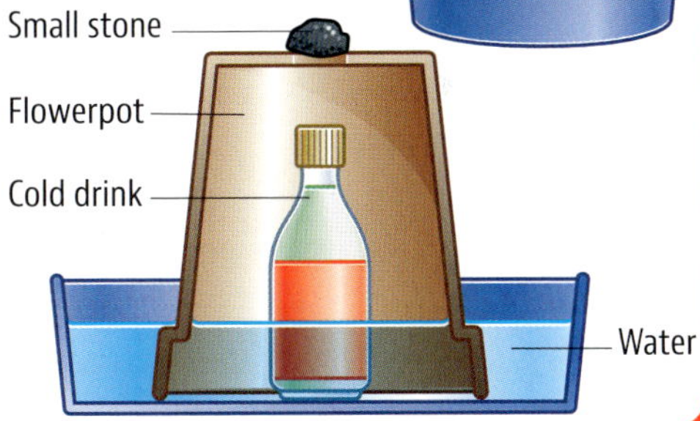

EXPANSION OF SOLIDS

When a solid is heated, the atoms or molecules from which it is composed vibrate more vigorously. Because these particles vibrate more, they take up more room. As a result, the solid expands to become bigger. The more it is heated, the larger it gets.

Expansion of solids, particularly metals, can cause problems. It can cause steel railroad lines to buckle and concrete roads to crack. In the Middle East, high daytime temperatures cause oil pipelines to lengthen, meaning they would buckle if the pipes had not been fitted with expansion loops. Even overhead power lines can droop dangerously low when the cables expand in the summer. However, the expansion of solids can also be useful. Some thermometers and thermostats rely on different expansion rates between two different metals bonded together (a bimetallic strip).

EXPANDING METALS

Different metals expand by different amounts when they are heated. The diagram below shows the amount of expansion when 16 ft (5 m) lengths of different metals are heated to 90°F (50°C).

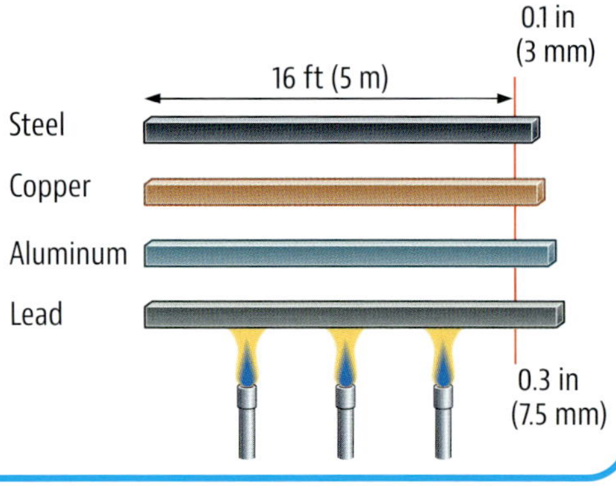

The metal girders on bridges expand in summer heat. Roller bearings allow the bridge to move without structural damage.

Moving Molecules

All solids are made up from atoms and molecules. The vibrations of these particles are what defines heat. Add more heat, and they vibrate more vigorously, raising the temperature of the solid. One consequence of this extra molecular motion is that the solid expands. A long bridge may expand by 3 ft (1 m) or more in the summer. The amount of expansion depends on what the bridge is made from. Steel, for example, usually expands more than concrete.

SCIENCE WORDS

Bimetallic strip: A strip that consists of two different types of metals that are bonded together. The metals have different rates of expansion, so the strip bends when heated.

Coefficient of linear expansion: A measure of how much the length of a solid object increases when it is heated.

Solid: A state of matter that keeps its own shape (unlike a gas or a liquid).

Measuring Expansion

A scientific measure of how much a solid increases in length when it is heated is its coefficient of linear expansion. This is defined as the proportion by which its original length increases when its temperature rises. Heated solids also increase in their volume, so they have in addition a coefficient of volume expansion (in the same way as liquids and gases). For example, think of a solid cube. When the cube is heated, each of its three dimensions increase in size. It is therefore not surprising that the coefficient of volume expansion is usually three times the value of the coefficient of linear expansion.

GAS CONTROL VALVE

When the valve is open, it admits gas to an oven's burners. As the oven heats up, a brass tube in the oven expands and lengthens. This action pulls a rod of Invar (an alloy that does not expand when heated) and closes the valve, which reduces the gas flow.

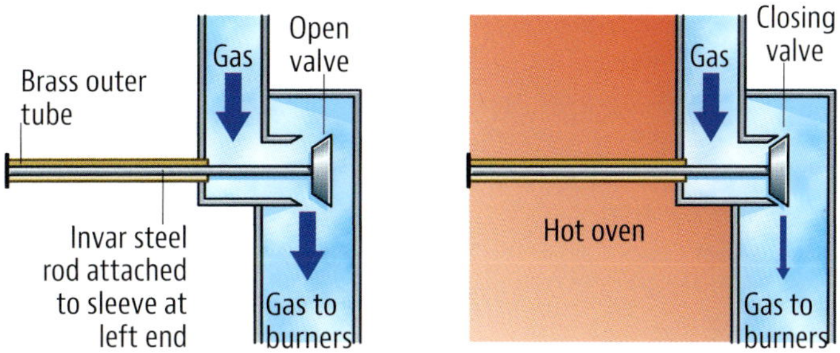

SOLIDS INTO LIQUIDS

The molecules or atoms that are present in a solid remain in a regular arrangement, which is what gives a solid its shape and strength. However, when heat energy is supplied to a solid it melts.

The effect of heat on a solid's molecules makes them vibrate faster. Eventually they break away from their regular positions and move about freely, just as in a liquid. In fact, the solid has become a liquid—it has melted. Every pure solid undergoes this change at a definite temperature, called its melting point.

The melting points of different substances span a wide range of temperatures. Ice is a familiar solid—it is the solid form of water. It melts at 32°F (0°C), and its melting point is used as the lower fixed point on the Celsius temperature scale. The metal mercury, on the other hand, melts at about −38°F (−39°C). That is why mercury is liquid at ordinary temperatures. Tungsten melts at the very high temperature of 6,170°F (3,410°C). It was once used to make the filaments of electric light bulbs. In fact, there are very few things that will not melt if they are made hot enough—even rocks and concrete will eventually melt.

SCIENCE WORDS

Bimetallic strip: A strip that consists of two different types of metals that are bonded together. The metals have different rates of expansion, so the strip bends when heated.
Coefficient of linear expansion: A measure of how much the length of a solid object increases when it is heated.
Solid: A state of matter that keeps its own shape (unlike a gas or a liquid).

Pressure also affects the melting point of a solid, although considerable pressure is needed to make a significant difference. The British physicist John Tyndall (1820–1893) devised a simple experiment to demonstrate what happens. He took a large block of ice, which is a solid, and supported it between two chairs. He then attached heavy weights to each end of a thin steel wire, and hung the wire around the ice block. The wire exerted greatest pressure on the ice immediately underneath it, and the ice melted to form liquid water. The wire sank a little way into the water before the ice refroze above it. In this way, with alternate melting and freezing, the wire gradually cut its way right through the block of ice.

We make snowballs by squeezing snow in our hands. The pressure melts some of the snow, which refreezes when the pressure is released, leaving a hard ball of snow. If the snow is too cold, squeezing it hard enough to melt it is difficult or impossible.

MELTING MOLECULES

When a solid is heated, its molecules move faster and away from their regular positions in the solid. Order changes into disorder. Eventually the solid melts and becomes a liquid.

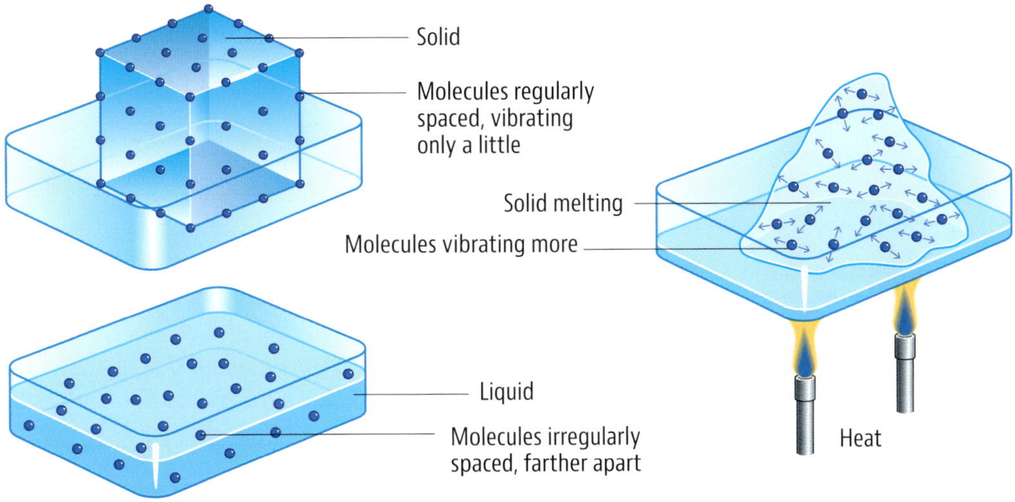

A stream of molten lava flows down the side of a volcano. The rocks deep underground take the form of magma, which is kept liquid by intense heat and comes to the surface as lava when volcanoes erupt.

Melting Moments

Most solids expand on melting because their molecules are farther apart in the liquid state. This effect is put to good use in an ingenious device that automatically opens the windows in a room or greenhouse when the temperature rises above a certain point. The device consists of a cylinder of wax with a tight-fitting piston. As the heat rises the wax melts and expands. This moves the piston, and the movement of the piston opens a window.

The chemical naphthalene also expands when it melts. This property can be used for keeping a constant temperature. A coil of wire carrying an electric current inside a can of naphthalene becomes warm and begins to melt the naphthalene. The expanding chemical presses a spring-loaded switch, which cuts off the electricity to the heating coil. The naphthalene cools and solidifies again. The switch restores the electric current, and the action is repeated.

CONDUCTION, CONVECTION, AND RADIATION

The handle of a metal spoon left in a hot drink quickly heats up and may become too hot to hold. The heat travels along the spoon through conduction. Materials that are poor conductors of heat are called insulators.

We all know that heat can travel easily through some substances. Others do not conduct heat so easily. The handle of a plastic spoon hardly gets warm if the spoon is left in a hot drink. How does heat travel in this way, and why are metals better heat conductors than plastics? The answers to these questions lies in the atomic or molecular structure of the materials.

A metal is made up from regularly spaced atoms that vibrate slightly around their usual positions. A "sea" of what are called free electrons occupies the spaces between the atoms. When a substance is heated, its atoms vibrate vigorously. The atoms at the hot end of a heated metal bar are the ones that vibrate the most. As they do this, they bump into their neighbors, increasing the neighbors' rate of vibration. Increased vibration results in higher temperature. In this way, the vibrations gradually increase along the length of the metal bar or spoon, and this is how the heat travels from the hot end toward the cool end. The atoms themselves do not move from their average positions. In a metal bar, free electrons moving rapidly along it carry most of the heat.

The internal structure of a plastic spoon is quite different. Plastics are composed of large molecules with no free electrons between them. The molecules do vibrate, and a similar ripple effect does conduct some heat along the spoon, but the process is very slow.

We therefore say that metals are good conductors of heat, while nonmetals such as wood and plastics are poor conductors, or insulators. A saucepan is made from metal so that it will rapidly conduct heat from a stove to cook the pan's contents, but the handle is usually wooden or plastic.

As well as heat, materials also conduct electricity. There are good electrical conductors and poor electrical conductors, which are also known as insulators. As with heat, most good conductors of electricity are metals, while nonmetals make up most of the poor conductors. This similarity is no coincidence. Both types of conduction depend

HEAT TRAVELS

If you put a metal rod in a pan of hot water, heat travels up the rod by conduction. Convection makes currents of water circulate up from the heated bottom of the pan, and heat also escapes from the electric stove by means of radiation. These are three ways that heat can travel.

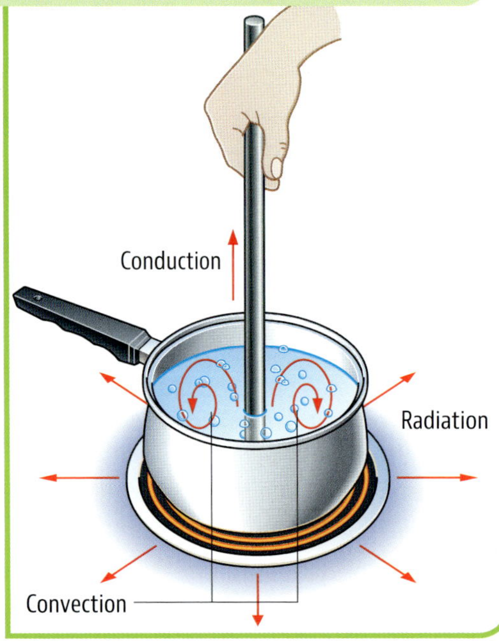

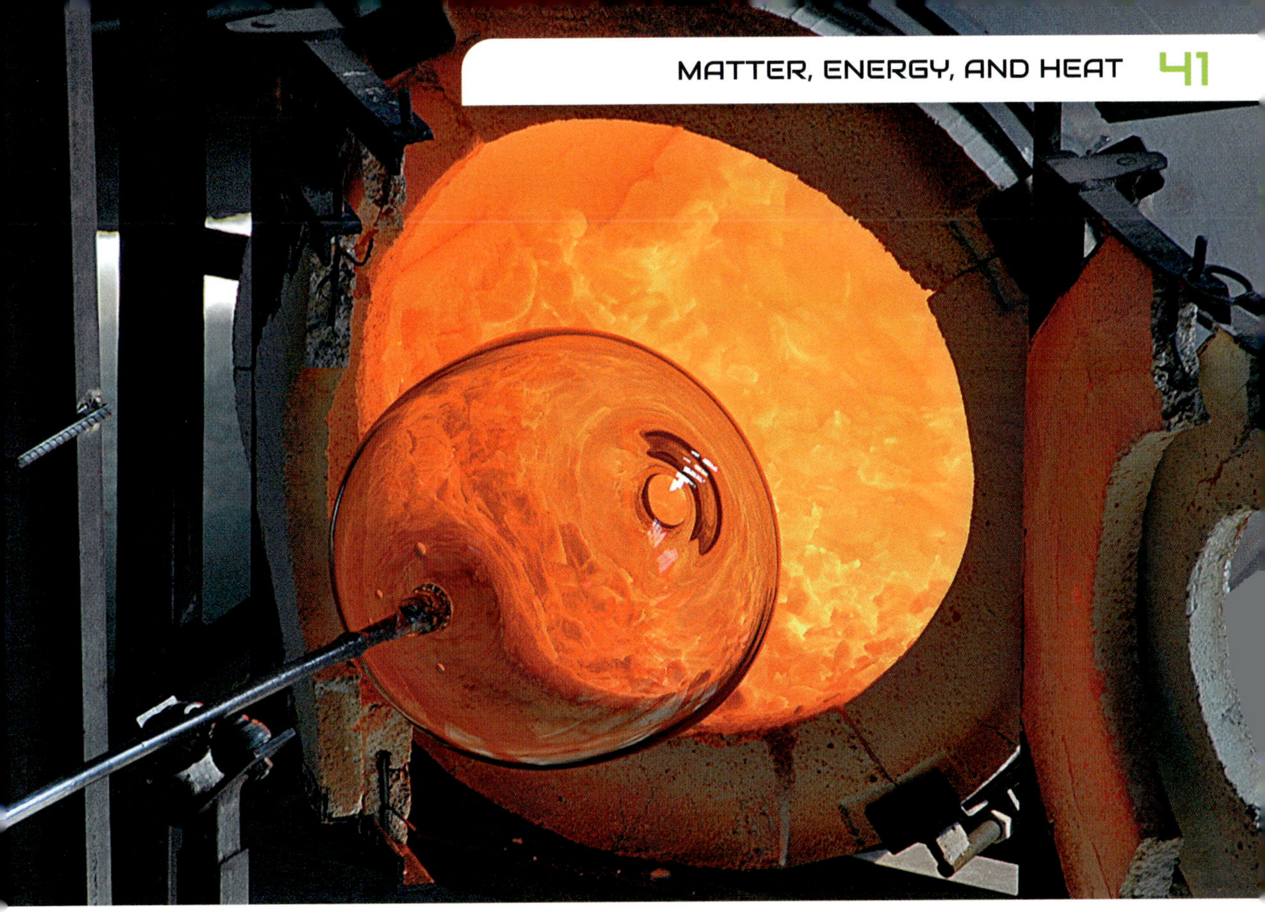

The temperature of a furnace has to be in excess of 1,650°F (900°C) to soften glass. The rod for turning the glass is made from a poor heat conductor to protect the glassblower's hands.

to some extent on the same property—the number of free electrons available in a material's structure. With electricity, the free electrons carry current through the material. With heat, the "hot" free electrons have a high kinetic energy, which they carry through the material to the colder part before giving it up in collisions with atoms. Insulators—both electrical and heat—do not have these free electrons to carry out the conduction function.

Rate of Heat Flow

Several things affect the rate at which heat is conducted by a substance. When one end of a block of material is heated, the rate of heat flow through the block depends on its length. The longer it is, the more slowly heat flows from one end to the other. The cross-sectional width of the block also matters—the greater this area, the faster heat will flow. The difference in temperature between the two ends of the block is also important. The greater this difference, the faster the heat will flow. Finally, there is a property of the material itself that is a measure of how well it conducts heat. This is the material's thermal

SCIENCE WORDS

Conduction: The process by which heat moves through a solid object. In electricity, conduction is the process by which an electric current moves through a substance.
Electron: A subatomic particle with a negative electric charge. Electrons surround the nucleus of an atom.
Kinetic energy: The energy an object possesses because it is moving.

conductivity. When heat is measured in joules, the units of thermal conductivity are watts per meter per K. The chart on page 44 compares the thermal conductivities of various substances. Silver is the best heat conductor of all, while the poorest are feathers and the vegetable fiber kapok. That is why these sorts of materials are used for heat insulation in bed covers and padded winter clothing.

Cork is a particularly poor conductor. You can feel this for yourself if you compare walking in bare feet on a ceramic tiled floor and walking on a floor covered with cork tiles. Cork feels much warmer because it does not easily conduct the heat away from your feet. For a similar reason, glass wool is used for insulating lofts and for wrapping water pipes to prevent them from freezing in winter.

Solid plastic foams, also called expanded plastics, are poor heat conductors (or good heat insulators). An example of this is expanded polystyrene (Styrofoam). You can hold a Styrofoam cup containing a hot drink because the plastic does not conduct the heat to your fingers. Foam plastics are sometimes pumped into the cavity walls of buildings. This keeps the building cool in the summer by keeping the Sun's heat out, and warm in the winter by stopping internal heat from escaping through the walls. Heat can also travel through fluids—liquids and gases–by conduction, although convection is the usual way heat travels in such substances.

Air has a thermal conductivity of 0.025 watts per meter per K, which is about the same as the conductivity of feathers. This means that air is a very poor heat conductor and a very good heat insulator. It explains why layers and loosely knitted clothing keeps you warm in winter. The air trapped in the materials acts as an additional insulator.

TRY THIS

Heat On the Move

Heat moves through solids by the process called conduction. In this experiment, you will see how well or badly three different solids conduct heat.

What to do

Stand a metal spoon, a plastic spoon, and a wooden spoon upright in a mug. Use a small blob of butter to stick a bead near the top of each spoon. Carefully pour very hot water into the mug and watch what happens.

Heat from the hot water gradually moves up the spoons. When the heat nears the end of a spoon, it melts the butter and the bead drops off. The first bead to drop off indicates the spoon that is the best conductor of heat. Which spoon was it? Which was next? You should find that the metal spoon was the better conductor. Poor conductors of heat are called heat insulators.

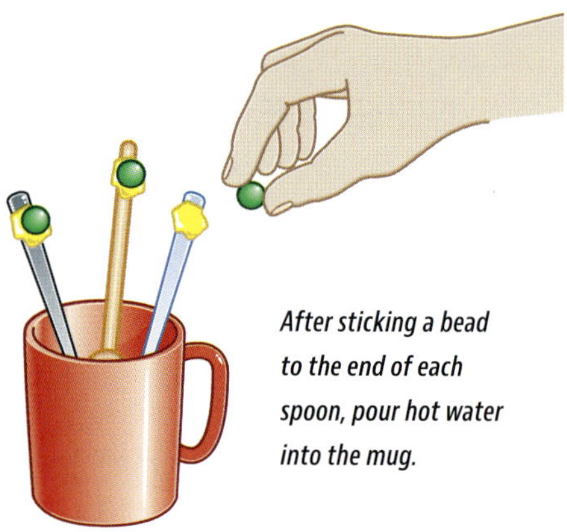

After sticking a bead to the end of each spoon, pour hot water into the mug.

Convection of Heat

Heat normally travels through a fluid—a gas or a liquid—by convection. In this process the material itself actually moves, which is why convection can't take place in solids. When a hot fluid moves, it takes heat with it and displaces any cold fluid in its way. Cold fluid also moves in to take the place of the rising hot fluid. In this way, the fluid circulates by creating a continuous convection current. The hot fluid is less dense than the cold fluid that surrounds it. As a result, it is more buoyant and tends to rise. Such natural convection currents have various uses. Soaring birds such as eagles make use of updrafts of warm air called thermals to help them fly effortlessly. Pilots of gliders and hang gliders also make use of thermals to keep their aircraft in the air without the use of motors.

Effect on the Weather

Large-scale convection currents in the atmosphere play a major part in creating and controlling the weather. As the air near the ground becomes warmer, it retains water vapor because warm air can hold more moisture than cold air. Eventually, the warm air rises, carrying the water vapor with it. When this air meets the colder air above, the water vapor condenses and forms tiny

The fur of a Polar bear keeps it warm in its Arctic habitat. Fur is a poor conductor of heat, so is a good insulator against the cold.

HEAT CONDUCTION IN ACTION

Heat travels through a solid by conduction. Imagine that the solid bar in this illustration has its lower end in a fire. The atoms at this hot end vibrate vigorously. These "hot" atoms joggle their neighbors, and they in turn joggle their neighbors, but not quite so much. The middle of the bar gets warm. Only at the cold end do the atoms vibrate normally. In a metal, heat is conducted mainly by free electrons moving between the atoms.

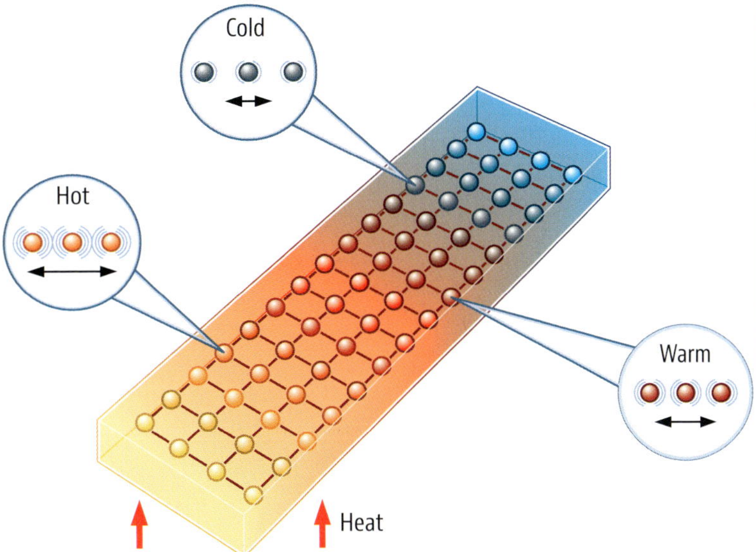

droplets of water. These masses of droplets form clouds called cumulus clouds. Convection currents also create winds and breezes. Near the coast in summer, the land warms up more quickly than the sea during the day. The warm air above the land rises, cools, and then sinks down again over the sea. This action sets up a circulating convection current with a gentle wind blowing onshore off the sea. This is called a sea breeze. The process is reversed at night, when the land cools down more rapidly than the sea. The result is a land breeze, which blows out to sea.

Liquid Convection

Warm water can also transfer heat by convection. It is less dense than surrounding colder water, so it rises, setting up a convection current. The principle is used in some hot-water heating systems in homes and offices. Hot water from a boiler rises and enters a cylindrical hot-water tank, where it transfers its heat to water that is circulating around the radiators.

Deep-Water Convection

On a much larger scale, convection takes place in the world's oceans. Water near the equator is warmer and less dense than water nearer the poles. This warmer water stays close to the surface and is displaced by colder water from polar regions.

The deep-water currents that result from this process shift huge volumes of water around the oceans. For example, every second around 1 billion gallons (5 million cu m) of water flows into the North Atlantic Ocean between Greenland and Iceland.

Forced Convection

Even when there are no density differences that will set up natural convection, blowers (for air) and pumps (for water) can circulate fluids to transfer heat. A hot-air heating

GOOD AND BAD CONDUCTORS

Thermal conductivities of common materials (in watts per meter per degree K).

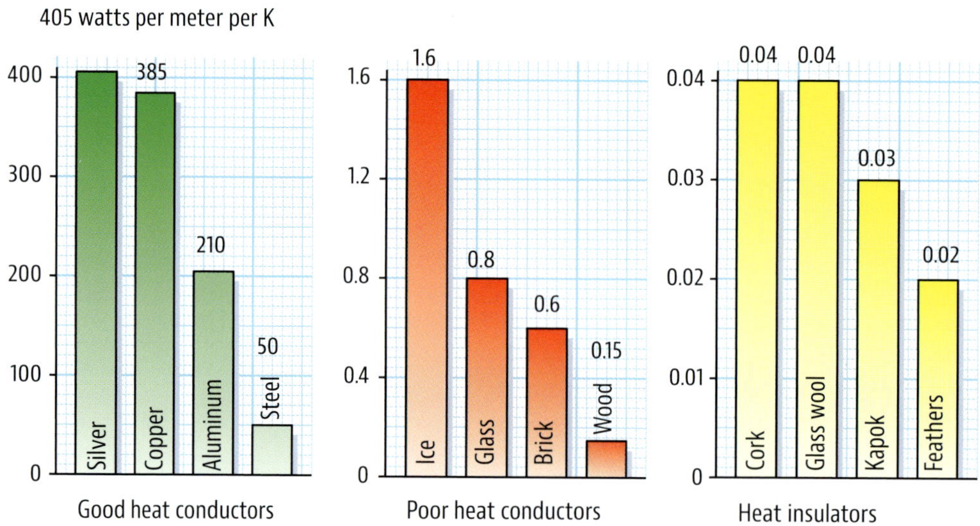

CLOUD FORMATION

Convection currents can cause clouds to form. Warm air, heated by the land, holds more water vapor than cold air. Convection makes this warm air rise. The water vapor cools and condenses as clouds.

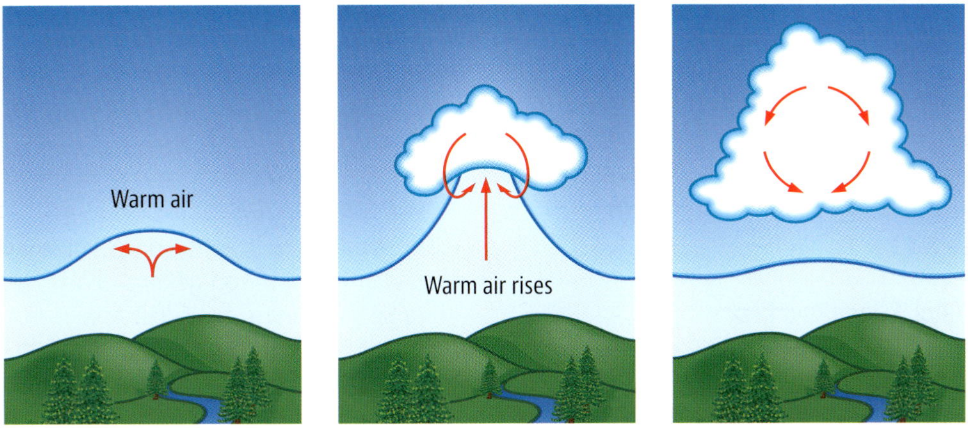

system has fans to blow air heated by a furnace along ducts into each room. In a car's cooling system, a pump circulates cooling water between the engine and the radiator.

Heat Radiation

In radiation, heat energy travels as electromagnetic waves. There are many kinds of these waves, including gamma rays, x-rays, microwaves, and radio waves. The most familiar are the waves of visible light—which enable us to see—and their invisible companions, ultraviolet light and infrared radiation. They all have different wavelengths, and they all travel at the speed of light. Infrared radiation possesses wavelengths that are slightly longer than the wavelengths of visible light.

Any object with a temperature above absolute zero gives off infrared radiation. This means that most objects emits some infrared, but the hotter the object, the more infrared it emits. You cannot see the radiation from a hot

Cumulus clouds (from the Latin word for "mass") form by the process described in the diagram above. The process requires the land to be warm, so these clouds usually indicate that the weather is fair.

radiator in a room, but you can feel it if you put your hand near the radiator. Another name for infrared is heat radiation. All the heat radiated by the Sun is infrared.

Detecting Radiation

Scientists use various instruments and techniques to detect electromagnetic radiation. Electronic circuits connected to an antenna can detect microwaves and radio waves. Gamma rays, x-rays, and visible light can be detected by photographic film, as in a camera or an x-ray machine, and special film can also detect ultraviolet light and infrared radiation. Astronomers use various electronic devices to study radiation that is emitted by heavenly bodies.

Another device for measuring infrared radiation is called a thermograph. This produces a picture of a hot object, with the temperatures of various areas represented as different colors or shades of gray. Thermographs are used widely in medicine for diagnosing tumors in the body. The skin over a tumor is always slightly warmer than the skin around it, and the tumor shows up as a light area on a thermograph.

Infrared radiation comes from a hot object's vibrating atoms or molecules. The more they vibrate, the more radiation they emit. But increasing the vibration of an object's atoms also increases its temperature. It therefore follows that the amount of infrared radiation an object emits increases as its temperature rises.

The wavelength of the radiation also depends on the temperature of the emitting object. The higher the temperature, in other words the greater the vibration, the shorter the wavelength. When an object is heated, it begins to emit more infrared. As its becomes hotter, the object becomes red. At this stage, its temperature is about 1,110–1,300°F (600–700°C). When it is even hotter, it turns orange (1,750°F/950°C), then yellow (2,000°F/1,100°C), and finally it becomes white hot at a temperature of about 2,550°F (1,400°C). At this stage the object will emit white light as well as infrared radiation.

SCIENCE WORDS

Convection: The usual way in which heat moves through a fluid (a liquid or gas) by setting up convection currents in the fluid.

Convection current: Movement of a fluid (a liquid or gas) that results when warm fluid rises and cold fluid flows in to take its place.

Thermal: A vertical convection current in a gas, usually air.

Thermal conductivity: A measure of the ability of a substance to conduct heat.

In a solar oven, a giant mirror focuses heat radiation from the Sun. It provides a source of cheap, clean heat for heating water and cooking. Large solar furnaces are used for melting metals.

Because of these changing colors it is possible to estimate a hot object's temperature just by looking at it. This is used by blacksmiths and other people who deal with very hot metals. It is also a method used by astronomers when they are studying the stars.

Although infrared radiation itself is invisible to the human eye, scientists can pick it up using light-detecting microchips that are sensitive to infrared. These have many practical uses. For example, infrared cameras on satellites orbiting the Earth, such as the Landsat series, can take infrared pictures of crops and other vegetation that are able to show up damaged or diseased plants among healthy ones.

Harnessing Radiant Energy

The largest source of radiant energy (light and infrared radiation) is the Sun. It lies 93 million miles (150 million km) from the Earth, and emits radiation in all directions across the vacuum of space. On the daylight side of the Earth, radiant energy from the Sun arrives at a rate of 1.4 kilowatts per square meter (kW per m2). One way of capturing some of this free energy is to use a large curved mirror to focus the Sun's radiation in a solar furnace.

The Sun's radiant energy is also used to make electricity. Solar cells capture the heat energy from the Sun and create electricity in solar panels. Different solar cells are made using different materials, depending on cost and what they are to be used for. The top layer of a solar cell has an antireflective glass coating that protects the semiconductors underneath. The sunlight is converted into electricity via layers made from silicon and phosphorus and silicon and boron. Solar cells are used in domestic solar panels, in spacecraft and satellites, and to generate electricity in engines. Solar power is an affordable, clean energy for the future.

TRY THIS

Underwater Fountain

There are underwater fountains called black smokers on the bottom of the deep oceans. They appear where hot liquids escape from cracks in the Earth's crust and rise into the cold water at the sea bottom. In this project, you will make your own underwater fountain.

What to do

Fill a bucket nearly to the top with cold water. Put a couple of marbles in a bottle as weights, or you could use any other small heavy objects such as a couple of steel nuts. Fill the bottle to the top with hot water and add several drops of food coloring. Cover the top of the bottle with plastic film and secure it with a rubber band. Stand the bottle at the bottom of the bucket of cold water. When the water has settled and become still, take a pencil and make a small hole in the plastic film with the point of the pencil, taking care not to disturb the water in the bucket too much. Carefully remove the pencil, and watch what happens.

The hot colored water is less dense than the cold water because the hot water molecules are moving faster and taking up more room. As a result, the hot water rises, just as a cork or piece of wood bobs up to the surface of water. This process is called convection and is what happens with the undersea black smokers.

Put two marbles in a small bottle of colored hot water. Put the bottle in a bucket of cold water and watch what happens.

RADIATORS AND ABSORBERS

Have you noticed that in warm weather you feel cooler in a white T-shirt than in a dark-colored one? Things that are black or dark-colored absorb heat more than things that are white or shiny. Black things are also better at emitting heat radiation.

Everything emits some radiation, but some things are better than others at emitting heat radiation. In general, black surfaces are better than surfaces that are white or shiny. A hot drink stays hot longer in a shiny metal mug than in a dark-colored china cup.

In addition, good emitters are good absorbers of heat radiation. The best absorber of all is a theoretical object called a black body, which absorbs all radiation falling on it. It is also the best emitter. Nobody has been able to make a perfect black body, but a mat black surface comes very close to perfection.

JAMES DEWAR

James Dewar was a Scottish scientist who lived and worked mainly in London, England. He is best known for inventing the vacuum bottle, or Dewar flask, in the early 1870s. This type of vacuum bottle is still used today. He used the Dewar flask during his researches into cryogenics (low temperatures), and by the year 1899 he had developed a large-scale method for liquefying hydrogen at a temperature of –423°F (–253°C). A year later, Dewar produced solid hydrogen at –434°F (–259°C). In 1889, he worked with Frederick Abel (1827–1902). Together they invented the propellant explosive cordite (based on guncotton), which was the first so-called smokeless ammunition. In 1904, Dewar was knighted for his work by King Edward VII.

For this reason, a car's radiator is generally painted mat black so that it emits the heat from the water circulating inside it. Solar water heaters on houses are thin tanks containing water. They are also painted mat black, but this time it is so that the tank absorbs as much of the Sun's heat radiation as possible in order to heat the water, which is then pumped into the house.

Stopping Heat Losses

A common need, both in science and in everyday life, is to prevent hot liquids from cooling down or cold liquids from warming up. The second of these problems faced the Scottish scientist James Dewar (1842–1923) when he was dealing with gases such as liquid nitrogen, which boils at –320.4°F (–195.8°C). His solution was the Dewar flask, or vacuum bottle, that we still use today for keeping hot drinks hot.

The vacuum bottle, also called a thermos bottle, works because it prevents the three methods of heat transfer—conduction,

The heat produced by the Sun travels as infrared radiation. You immediately feel the loss of this heat if the Sun goes behind a cloud or dips below the horizon at sunset.

convection, and radiation. The bottle consists of two thin-walled glass bottles, one inside the other, with all the air removed from the space between them. This vacuum prevents heat from traveling by conduction because there is nothing for the heat to travel in. It also prevents heat from being transferred by convection because there is no fluid to circulate and carry the heat (see diagram, right).

In addition, the glass walls of the double bottle are covered with a layer of silver, like a mirror. This silvered surface minimizes heat loss from radiation because shiny objects are poor emitters of radiant heat.

Trapping Heat

When an object absorbs heat radiation it becomes hotter than its surroundings. Any object that is hotter than its surroundings gives off heat. The heat emitted has a longer wavelength than the heat it absorbs. This is used to good effect in a greenhouse, which traps the heat. The soil and plants inside a greenhouse absorb radiation from the sunlight that passes through the glass, making them warmer. The plants then reradiate the heat as longer-wavelength infrared radiation. However, these longer waves cannot pass through glass. As a result, the heat from the plants is trapped inside the greenhouse. The temperature inside the greenhouse may be up to 27 to 36°F (15 to 20°C) warmer than the temperature outside.

VACUUM BOTTLE

The vacuum bottle, or Dewar flask, is designed to prevent the loss of heat through conduction, convection, and radiation. The vacuum prevents heat loss by conduction or convection, and the silver mirror minimizes loss through radiation. The stopper prevents heat loss through evaporation.

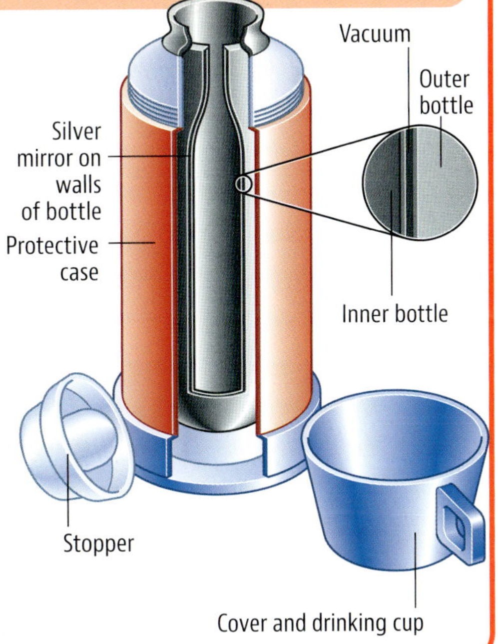

EFFECTS OF COOLING

Here we look at how cooling affects the arrangement of atoms and molecules.

Cooling a substance has the effect of slowing down the movement of atoms and molecules. When a gas is cooled, its molecules move more slowly and do not travel so far. This has the effect of reducing the pressure of the gas. If it is made even cooler, the molecules begin to behave like those in a liquid. The gas (or vapor) will have condensed into a liquid. That's what happens to the steam coming from the spout of a kettle of boiling water. In the cooler air, the steam condenses to form tiny droplets of water. They are the white "steam" that we see; like clouds in the sky, steam consists of water droplets.

As a gas cools, its volume decreases. At constant pressure all gases "shrink" by 1/273 of their volume at 32°F (0°C) for every degree fall in temperature. This relationship between the volume of a gas and temperature is called Charles's Law, after the French scientist Jacques Charles (1746–1823) who discovered it (see diagram on page 30). The law is usually stated in this form: the volume of a fixed mass of gas at constant pressure is directly proportional to its absolute temperature.

SCIENCE WORDS

Condensation: The process by which a gas or vapor changes into a liquid. The liquid formed is also sometimes called condensation.

Superconductor: A substance that shows no resistance to the passage of electric current; some metals do this when cooled close to absolute zero (−459.67°F/−273.15°C).

Liquid Gases

If the pressure on a gas is increased sufficiently, it turns into a liquid. Or at least this is true of many gases. There are some, however, that have to be cooled as well as being put under pressure before they will liquefy. The temperature below which they must be cooled is called the critical temperature. For example, carbon dioxide gas cannot be liquefied by pressure alone unless it is cooled below 88.3°F (31.3°C), which is its critical temperature. Some gases have to be made extremely cold to liquefy them.

COOLING SEQUENCE

Cool a gas sufficiently, and it condenses into a liquid. Cool a liquid, and it eventually freezes to form a solid.

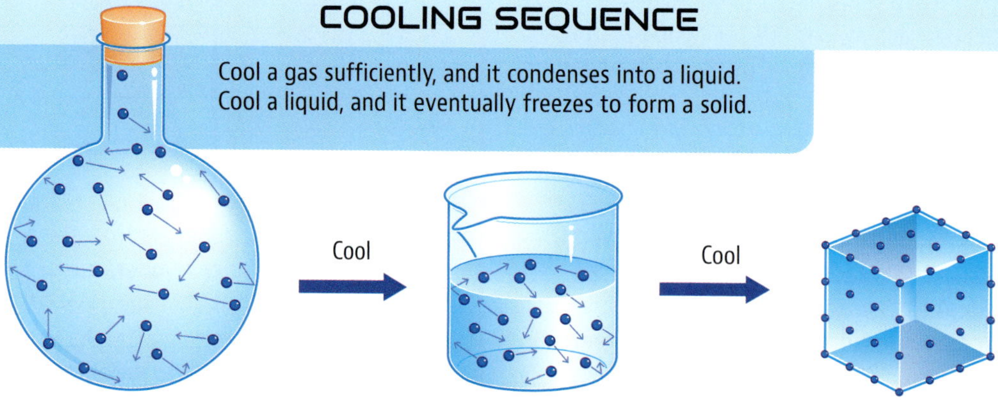

Gas: molecules moving about violently

Liquid: molecules moving/vibrating

Solid: molecules in crystal lattice, vibrating only a little

At ordinary pressure, oxygen liquefies at −297.4°F (−183°C), and liquid nitrogen forms at −320.4°F (−195.8°C). Liquid air also has a temperature of about −321°F (−196°C). Liquefied gases such as liquid helium, which is the coldest, are used to surround and cool large computer memories and also big magnets used in medical scanning machines. Liquid hydrogen is used as a rocket fuel, and liquid nitrogen is employed as a refrigerant and for liquefying natural gas.

Cooling Liquids

If cooling a gas turns it into a liquid, cooling a liquid turns it into a solid. The temperature at which this occurs is called the freezing point of the liquid. It is exactly the same as the melting point of the solid.

Sometimes you can cool a liquid below its melting point without it freezing. This is called supercooling. If the temperature of the supercooled liquid rises slightly, the solid and liquid forms exist together for a while until the whole lot solidifies.

Cooling Solids

As a liquid cools, it slowly crystallizes to form a solid. The nature of the crystals depend on how fast the liquid cools. If it cools rapidly, only small crystals have time to form. But if the liquid cools very slowly, there is time for large crystals to grow. These effects are best seen with minerals, large crystals of which form only if cooling has been slow. Most solids shrink in size as they get colder, but ice actually expands, by as much as 10 percent. This can have disastrous effects on water pipes and the engine blocks of automobiles, which may burst if the water in them freezes.

Metals may also have unusual properties at low temperatures. The crystal structure of some metals changes. The crystals grow larger, and the metal loses strength, becoming brittle and snapping easily. At very low temperatures, close to absolute zero, some metals lose their electrical resistance. There seems to be nothing to stem the flow of electricity through them, and they become what are called superconductors. Scientists continue to search for materials that are superconductors at higher temperatures. They want to find metals that become superconducting at room temperature, which they believe will help to achieve faster, more powerful supercomputers, new forms of transportation systems, and improvements in global energy-producing technology.

TRY THIS

Lifting the Ice

This project can be made into a puzzle. How can you lift a cube of ice without touching it with your fingers? Simple—make it even colder!

What to do

Half fill a glass with water and float an ice cube on the surface. Lay one end of a piece of string across the ice cube, and sprinkle salt over the string. Wait for 30 seconds, then gently lift the string. The ice cube will have frozen onto the string and can be lifted with it.

Pure water freezes at a temperature of 32°F (0°C), but salt water freezes at a much lower temperature. This is why it has to be extremely cold for the oceans to freeze. When you sprinkle salt on the string lying across the ice cube, the freezing point is lowered. At first, the ice will melt a little, but the coldness of the ice cube then causes the salt water to freeze again, locking the string onto the cube as if it were glued on.

PRESSURE ON GASES

Gases are the easiest states of matter to compress—increase the pressure on a gas, and its volume gets smaller. This can be an extremely useful property.

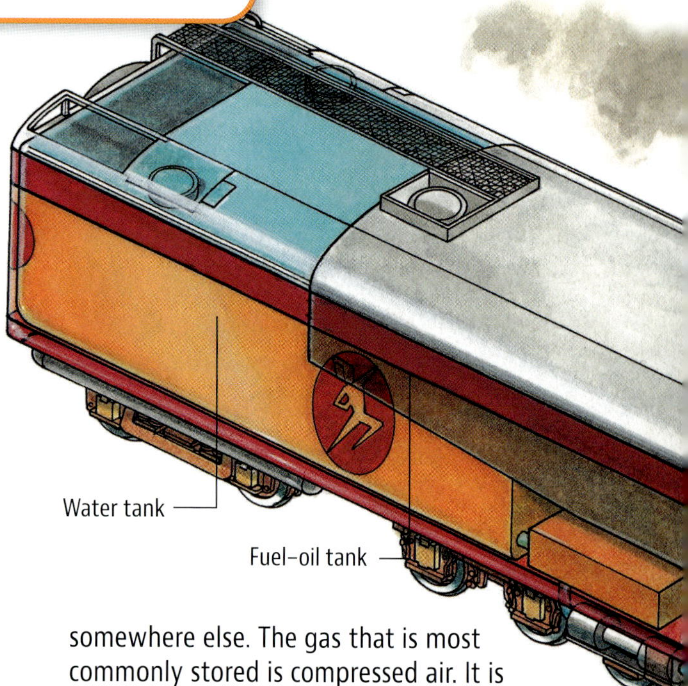

Compressing a gas requires an input of energy. The resulting compressed gas is therefore a store of energy that is capable of doing work. Scientists and engineers have devised many ways of using the energy of gases under pressure. It is important to remember that, scientifically, a vapor is a gas. Common vapors include steam and the mixture of gasoline and air that provides the source of energy in an automobile engine.

Compressing Gases

To make use of compressed gases, they must either be compressed where they are to be used, or stored under pressure for use somewhere else. The gas that is most commonly stored is compressed air. It is produced by a machine called a compressor. Most compressors consist of an engine or a motor that works a piston. The piston moves rapidly back and forth in a cylinder to compress air, which is stored in a strong metal tank or in cast-iron bottles called cylinders.

Cylinders of compressed air have many uses, from blowing up balloons for a party to working a spraygun for repainting an automobile. You may have seen portable compressors on trailers where workers are breaking up the concrete of roads or old buildings. The jackhammers they use are powered by compressed air. Another type of compressor has a rotary motion and works a bit like a water pump. The fastest types are equipped with blades like the blades of a turbine. Every jet engine has a compressor like this to compress the air used to burn the vaporized fuel in the engine.

SIMPLE AIR COMPRESSOR

A bicycle pump is an air compressor. When the handle is pulled back **(a)**, air moves past the piston into the barrel. When the handle is pushed in **(b)**, the piston fits tightly and compresses the air.

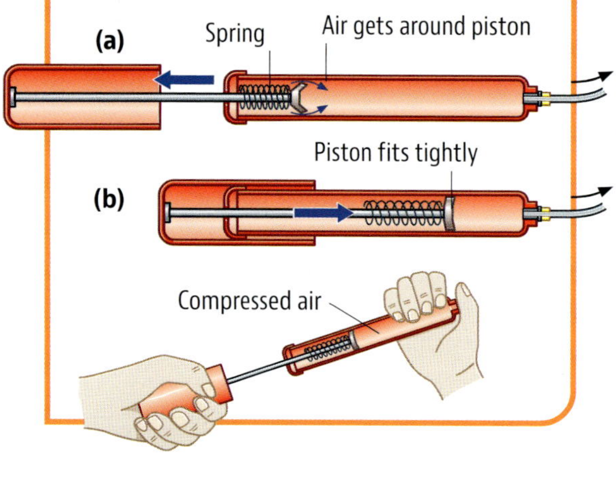

The Steam Engine

The first machine to exploit gas pressure was the steam engine. Invented in the 18th

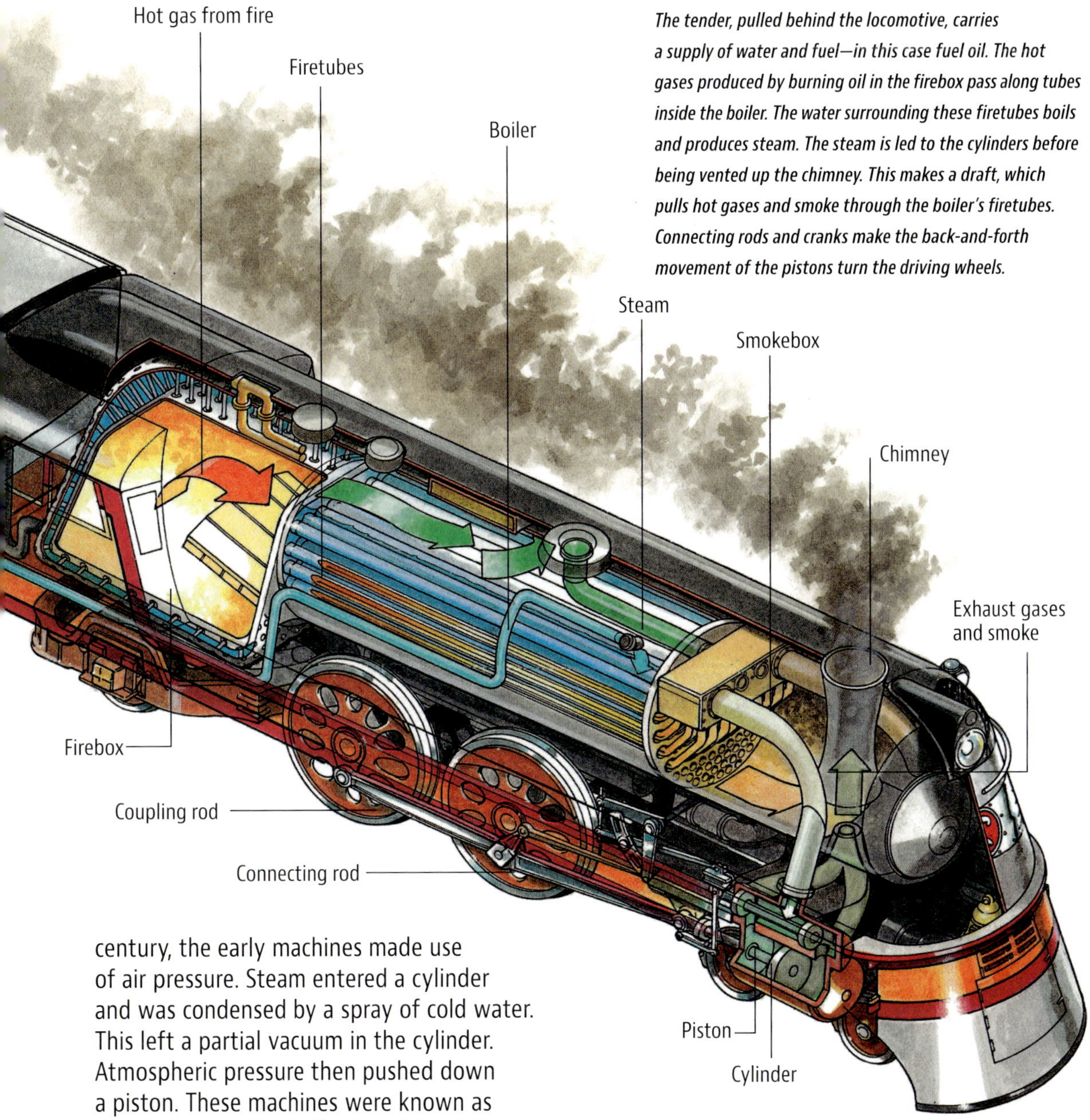

The tender, pulled behind the locomotive, carries a supply of water and fuel—in this case fuel oil. The hot gases produced by burning oil in the firebox pass along tubes inside the boiler. The water surrounding these firetubes boils and produces steam. The steam is led to the cylinders before being vented up the chimney. This makes a draft, which pulls hot gases and smoke through the boiler's firetubes. Connecting rods and cranks make the back-and-forth movement of the pistons turn the driving wheels.

century, the early machines made use of air pressure. Steam entered a cylinder and was condensed by a spray of cold water. This left a partial vacuum in the cylinder. Atmospheric pressure then pushed down a piston. These machines were known as atmospheric engines. The Scottish engineer James Watt (1736–1819) then began to build engines that used the pressure of steam to move the piston. This was the basic principle used for all steam engines that were to follow.

The invention of the steam engine changed the world. It ushered in the Industrial Revolution (c.1760–c.1840), a period of great technological and social change in Britain. Stationary steam engines drove spinning machines and looms in textile factories, and portable engines on wheels were employed for plowing. Steam engines powered boats, road carriages, and eventually locomotives for railroads. The key parts of a steam locomotive are a fire, which heats a boiler for making steam, and a cylinder mechanism that uses steam pressure to produce motion. The fuel for the fire may be wood, coal, or oil. The heat from the fire passes along firetubes that run the length of the boiler and are surrounded by water. The steam, under high pressure, then moves a piston back and forth in the cylinders. Systems of valves open and close to let the steam into the cylinders, and out again when it has done its work. The waste steam passes up a chimney, along with the smoke from the fire. In a steam engine, the fuel is burned outside the engine itself—it is an external- combustion engine. Toward the end of the 19th century, inventors in Germany and elsewhere began to make engines that burned fuel inside them—these were the internal-combustion engines.

Internal-Combustion Engines

The earliest type of internal-combustion engine used flammable gas, such as coal gas, and later a mixture of gasoline and air, as fuel. The fuel and air were compressed inside the cylinders of the engine, and the rapidly expanding hot gas moved the pistons within the cylinders. An electric spark from a spark plug ignited the gasoline vapor.

SCIENCE WORDS

Compressor: A machine for compressing a gas, or putting a gas under high pressure.
Partial vacuum: An area of low atmospheric pressure, especially one in which most of the air has been pumped out.
Vapor: Another name for the gas that forms when a liquid boils or evaporates.

A jackhammer is powered by compressed air. The air is released in rapid bursts, causing a piston to hit the top end of the cutting chisel. It is not a new idea. A French engineer invented it in 1861 to drill the first tunnel through the Alps between France and Italy.

This is still, in principle, how most of the world's automobile engines work.

In 1885, the German engineer Rudolf Diesel (1858–1913) began to develop a compression ignition engine. In the diesel engine, as it came to be called, air was compressed until it was so hot that it ignited the vaporized fuel without the need for spark plugs. Diesel engines are commonly used in trucks, trains, and automobiles. However, although diesel engines produce less CO_2 emissions than petrol engines, they produce high levels of particulates, or tiny pieces of soot, that are just as bad for the health.

A third type of internal-combustion engine has no pistons at all. It uses the hot gases from burning fuel to turn the blades of a turbine and for this reason it is called a gas-turbine engine. The blades of a compressor are mounted on the same shaft as the turbine blades. It is located in front of the turbine, and its job is to compress air for feeding into the combustion chamber. When they are first started, these engines need an electric motor to spin the shaft and get the compressor working before fuel is injected and ignited. A jet of hot gases leaves the rear of the engine, giving it its more common name—the jet engine. There are various types of jet engine, used for both military and civil aircraft.

Gas-turbine engines power some modern high-speed railroad locomotives. They are also used in small power plants to drive electricity generators, and as standby engines in large power plants for use in the event of a steam-turbine breakdown.

Rockets and Propellants

Rocket engines are, in some ways, similar to jet engines. They also burn fuel in a combustion chamber, and the resulting hot gases expand through a nozzle at the rear, creating the thrust that propels the rocket. However, it is not the jet of hot gases pushing against the air that creates the thrust of a rocket engine. It is the reaction force against the front of the combustion chamber that gives the forward "push."

Internal combustion engines and steam engines use the oxygen in the air to burn their fuel. A rocket engine is different because rockets carry their own oxygen supply. A liquid-fueled rocket, such as the giant Saturn V that launched Apollo 11 in 1969, has tanks of liquid oxygen and liquid hydrogen as fuel. It takes time to prepare a liquid-fueled rocket for launching, but solid-fueled rockets are ready to be launched immediately. The simplest form of solid-fuel rocket burns gunpowder. They power military missiles and act as launch boosters for larger rockets, and because they have their own supply of oxygen, all rockets work in the airless conditions of outer space.

The propellants inside a bullet cartridge are similar to solid rocket fuel. The cartridge has three parts: the primer, the gunpowder propellant, and the bullet (the pointed part of the cartridge that exits through the gun barrel). When the trigger is pulled, a firing pin explodes a small charge in the primer, which ignites the propellant in the casing. The large volume of hot expanding gases this produces propels the bullet with great force out of the gun barrel.

TRY THIS

Crush a Bottle

A simple way to demonstrate the force of air pressure around us uses atmospheric pressure to crush a bottle.

What to do

Unscrew the cap from an empty plastic soda bottle and carefully add about half a cupful of hot water from the faucet. Leave the bottle to stand for a couple of minutes, then pour out the water. Quickly screw the cap back on tightly, and watch what happens. It is as if an invisible pair of hands is slowly crushing the bottle.

After the cap is screwed on, the bottle starts to cool down. As it cools, hot water vapor inside the bottle condenses and forms small droplets of water. They take up far less room than the water vapor, but no air can enter the bottle to fill up the empty space. As a result, a partial vacuum forms inside the bottle. The pressure of the air outside the bottle then crushes it (because there is not enough air inside the bottle to push back). Normally, even an empty bottle is full of air.

Let the bottle and the hot water in it stand for a couple of minutes before you screw on the cap.

PRESSURE ON LIQUIDS

Liquids are very difficult to compress. The pressure applied at one place in a liquid is transmitted immediately to every other part of it. This principle is used in hydraulic machinery, from large presses for stamping out automobile bodies to bulldozers and diggers.

The molecules in a gas are relatively far apart. When pressure is applied to a gas, the molecules are squeezed closer together, and the volume of gas becomes smaller. The molecules in a liquid are very hard to squeeze closer together. As a result, a liquid cannot really be compressed. Applying pressure can, however, move a liquid. This is what a pump does. The pump on a fire truck produces a jet of water that can reach the roof of a building.

Pressure Units

Pressure is a force acting over an area. It is equal to the total force divided by the area. The scientific unit of pressure is called the pascal, named after a French physicist who studied the subject. In customary units, pressure is usually given in lb per sq in (which is sometimes abbreviated to psi). A single lb per sq in is equal to nearly 7,000 pascals. There are various other pressure units. The millibar is used in weather forecasting to express the pressure of the atmosphere. One millibar equals 100 pascals. The word "atmosphere" is also the name for a unit of pressure. It is equal to the average pressure of the Earth's atmosphere at sea level and is about 100,000 pascals or 14.7 lb per sq in.

SCIENCE WORDS

Hydraulic press: A machine that uses the pressure of a liquid to "magnify" a force. A small force on a small piston produces a much larger force on a larger piston.
Molecule: A combination of at least two atoms that forms the smallest unit of a chemical element or compound.
pascal: Standard scientific unit of pressure.

Hydraulic Press

A hydraulic press has a small piston and a large piston that are connected by a pipe. The mechanism is filled with an oily liquid. Because pressure is transmitted equally throughout a liquid, a small force on the small piston results in a greater force on the large piston. For example, if the large piston has ten times the area of the small piston, a weight of 2.2 lb (1 kg) placed on the small piston will raise a weight of 22 lb (10 kg) placed on the large piston.

However, the large piston will move up only a tenth as far as the small piston moves down. For this reason, a practical hydraulic press has an arrangement of valves that allows the small piston to be pumped down repeatedly to raise a heavy weight. A common application for this is in the braking system

To reach the top of a burning building, the water in a firefighter's hose has to be supplied at high pressure. A pump in the fire truck provides the pressure.

HYDRAULIC PRESS

Pressing down on the small piston makes the large piston move up. The pressure is the same on both pistons, but the force is not the same. A small force on the small piston creates a large force on the large piston.

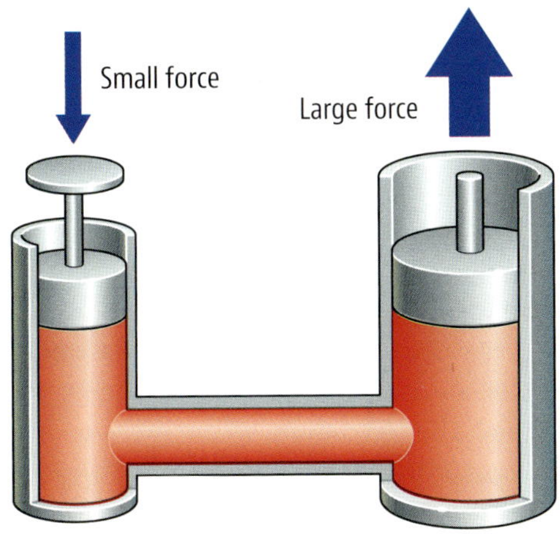

of automobiles and trucks. When the driver presses on the brake pedal, it pushes a piston in a large cylinder. Pipes connect this cylinder with smaller cylinders positioned on each wheel of the vehicle. Hydraulic fluid in the pipes transmits the pressure to the small cylinders, where the movement of their pistons applies the brakes. Because of the magnifying effect of using small and large cylinders, a fairly small force from the driver's foot then creates a large force, which applies the brakes.

Dump trucks and bulldozers use hydraulic pressure. Even large aircraft use hydraulics to operate the undercarriage and wing flaps. Most of these actions do not rely on the movement of a piston in a small cylinder to create the fluid pressure. Instead, rotary pumps pressurize the hydraulic fluid, which is routed by a system of valves.

All the movements in the arm and bucket of a digger are produced by hydraulics. Notice the piston-and-cylinder arrangement at each of the pivots. A pump pressurizes the fluid inside the pistons and in the pipework connecting them.

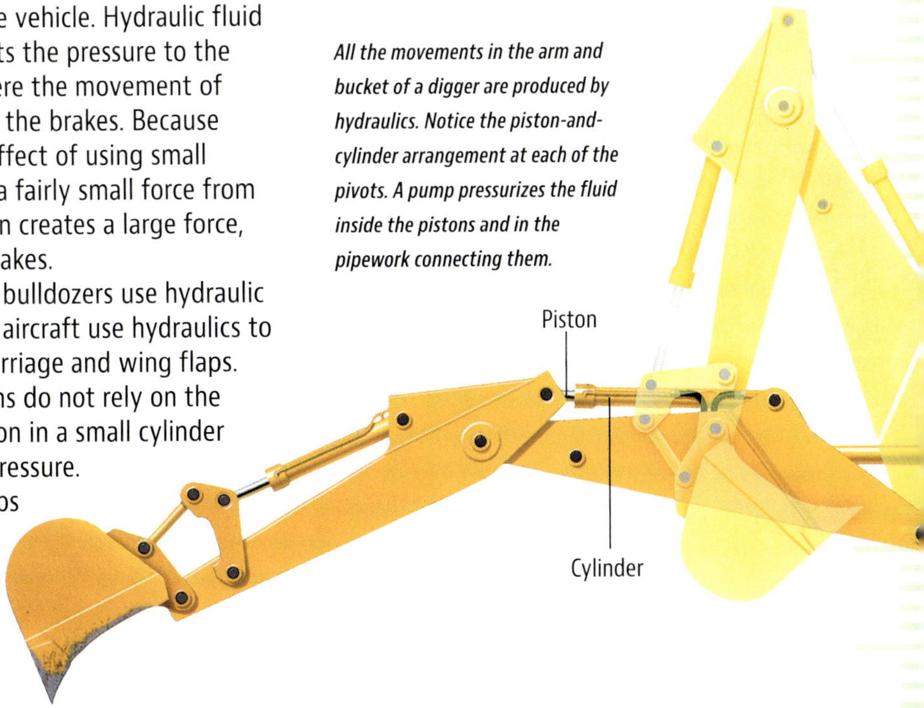

FLOWING FLUIDS

Although there are many differences between gases and liquids, they can sometimes be alike. In these cases, scientists use the word "fluid" for both gases and liquids.

The lemon shark is one of the fastest fish in the sea. Its body is perfectly streamlined so that it moves through the water effortlessly. The water nearest the shark's body moves along with it. As a result, the tiny striped pilot fish can swim along with the shark without getting left behind.

Engineers need to know about the way things move through fluids. A speedboat slicing through the water, a jet plane flying through the air, and a supersonic bullet speeding to its target all have to be designed to push their way through a fluid. But it is difficult to study objects moving through fluids. Instead, engineers usually observe the fluid flowing past the object. The results are much the same.

All surfaces that make contact with each other experience friction, and objects moving in fluids are no exception. Friction opposes the movement, so it has the effect of slowing down a moving object. It also generates heat. A bullet is very hot when it reaches its target, and the surface of a meteorite burns away through friction when it arrives from space and plunges into Earth's atmosphere at many miles per second.

One way of reducing the effect of friction on fast-moving objects is to design their shape so that friction is reduced to a minimum. A rifle bullet is streamlined. Round bullets or square ones would not travel straight through the air—they would tumble and probably miss their target. Streamlining accounts for the curved shapes of ship's hulls and high-speed airplanes. In the natural world, fish such as sharks and birds such as swallows also have streamlined shapes.

Smooth or Turbulent?

Streamlines are the paths followed by all the particles of a fluid as they pass a particular point. It is as if the fluid consists of parallel layers—in fact, streamlined flow is also called laminar flow (from the Latin word for "layer"). The fluid as a whole keeps on moving smoothly in the same direction.

If an object moving in a fluid (or around which a fluid is flowing) does not have a streamlined shape, eddies and whirls begin to form in the fluid. This is called turbulent flow. It can be seen, for example, in the water at the foot of a waterfall.

Turbulence can also affect fluids flowing through pipes. That is why pipelines are made smooth on the inside, with no obstructions or sharp bends. In this way the fluid flows as quickly as possible. The rate of flow equals the speed of the fluid multiplied by the cross-sectional area of the pipe. If the speed cannot be kept as fast as possible, the only way to increase the rate of flow is to make the pipe wider—which is a very expensive option.

Changing Speed of Flow

If a pipe includes a section that is narrower than the rest of it, the speed of flow is obviously affected. The speed increases in the narrow part, to maintain the same overall rate of flow through the whole pipe.

At the same time, the pressure of the fluid falls in the narrow part. The actual pressure change can be calculated using a complicated mathematical expression called Bernoulli's equation, which also takes into account the fluid's density. The equation makes some useful predictions about fluid pressure. It suggests that the pressure in a fluid depends on its depth. That is why a dam is built from thicker concrete near the bottom, where the pressure of water behind it is greatest.

Pressure also increases with depth in the sea, which is why the hulls of submarines have to be made very strong. If they were not, they would be crushed by the pressure of the water. There is a simple experiment to demonstrate this effect. Take an empty 70 fl oz (2 liter) plastic drink bottle and make three or four holes one above the other down one side. Put a single strip of tape over all the holes, then fill the bottle with water. Place the bottle in the sink and pull off the tape. The water will jet out of the holes, but which stream goes farthest? It is the one near the bottom of the bottle, where the pressure is the greatest.

There are a number of practical applications of Bernoulli's equation, which predicts that a reduction in pressure of a flowing fluid causes an increase in the speed of flow. In a chemist's

MOVING THROUGH FLUID

When a rectangular block moves through a fluid **(a)**, it creates great turbulence. A sphere **(b)** creates less turbulence, but the best shape of all is a teardrop, which creates no turbulence **(c)**.

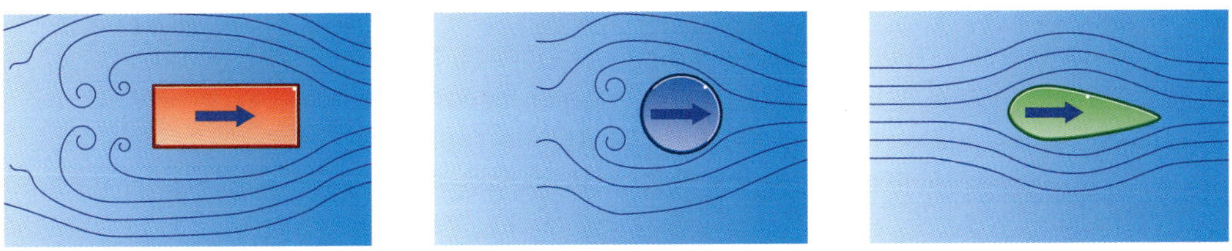

(a) (b) (c)

Bunsen burner or in a plumber's blowtorch, for example, gas passes through a small jet before it burns at the end of a tube. This speeds up the flow of gas and lowers the pressure, which sucks in air through the holes in the sleeve around the end of the tube. The air mixes with the gas and produces a hotter flame than if the gas had burned on its own.

Drag and Lift

It takes force to push an object through a fluid—jet airliners need very powerful engines. They are designed to move the airplane forward through the air. The resistance to movement is called drag, and streamlining is one way of keeping drag as low as possible. But how does an airplane fly in the first place? The answer has to do with the design of the wings, which move the plane upward.

Most airplane wings are slightly more curved on the upper surface than on the lower surface (see the diagram, above right). This shape is called an airfoil. As the wing moves through the air, the airflow is diverted. Some air passes under the airfoil, and some flows across the upper surface. As the plane flies forward, the top curve of the wing lowers the air pressure, moving the plane upward. The air pressure is higher beneath the wing than it is above it. The difference in air pressure causes a difference in air speed. The air molecules over the top of the wing arrive at the rear of the wing faster than those moving under the airfoil. Both are speeding downward, and this provides an upward force called lift. Lift is what keeps the plane in the air.

AIRFOIL SURFACES

The airfoil that forms an airplane's wing is slightly more curved over the top surface than underneath. This results in an upward force called lift, which keeps the airplane in the air.

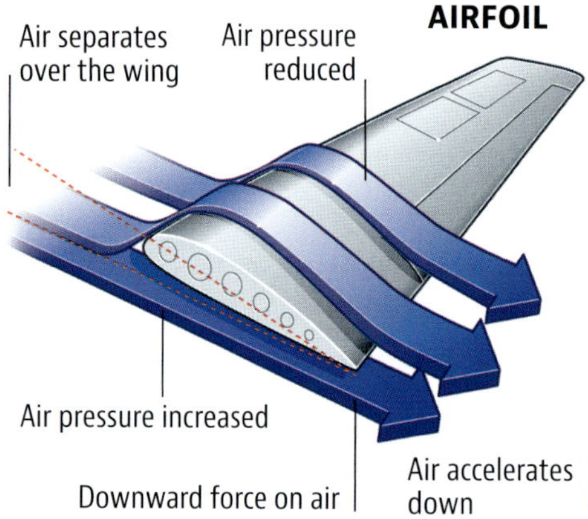

SCIENCE WORDS

Airfoil: The cross-sectional shape of an airplane wing, curved more over the top than beneath.
Lift: A mechanical force generated by a solid object moving through a fluid.
Turbulence: The irregular flow of a fluid moving around an object. It is reduced by streamlining.

To achieve maximum lift, the whole wing is angled upward slightly. This angle is called the angle of attack. If the wing is angled too much, the airflow ceases to be streamlined and turbulence sets in behind the wing. At too steep an angle, the drag increases and the lift decreases rapidly. This sudden loss of lift is called a "stall." Wind flaps are used to decrease stalling speed as an aircraft comes in to land.

Getting It Right

Cars, airplanes, and the hulls of ships are all designed to be streamlined to reduce drag and improve fuel efficiency. Air turbulence makes engines work harder and use more fuel. Engineers use models to test streamlining. Models of ships' hulls are tested in a long narrow tank of water called a ship tank. Models of cars and planes are tested in wind tunnels. The model is kept stationary, and air is blown past it at high speed by large fans. Smoke is often introduced into the airflow to highlight the streamlines around the model.

A jet plane coming in to land. Extendable flaps on the trailing edges of the wings increase the drag and decrease the stalling speed, slowing the aircraft safely as it approaches the runway.

TRY THIS

Bending the Draft

If you blow at a piece of paper held vertically, the draft from your breath will bend the paper. If you put an obstruction between you and the paper, you would expect it to block the draft so that the paper does not bend, but it all depends on the obstruction.

What to do

Cut a strip of paper about 4 in (10 cm) long and 0.5 in (1.3 cm) wide. Fold up one end so that it looks like a long letter "L." Use tape to fasten the paper to a worktop, as shown below. Blow at the paper. You will see it bends easily. Now place a bottle in front of the "L" to act as an obstruction. Blow in the direction of the paper. Try blowing quickly and slowly. When you get the blowing just right, the paper will bend just as it did before. What has happened?

The bottle has bent the air. This is because moving air tends to follow a curved surface. The bottle splits the air and carries a draft around its curved sides. The two drafts recombine on the other side of the bottle and continue forward to the paper.

The bottle bends the draft around its curved surface.

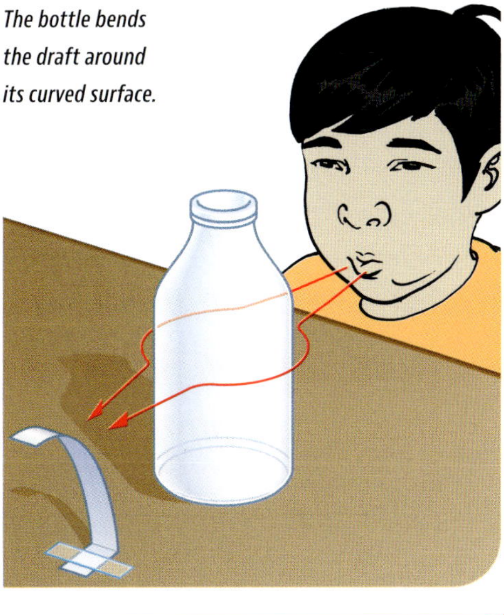

GLOSSARY

Absolute temperature scale The temperature scale that begins at absolute zero. It is also called the Kelvin temperature scale.

Airfoil The cross-sectional shape of an airplane wing, curved more over the top than beneath. Moving through the air, an airfoil produces lift.

Atmospheric pressure The pressure of the Earth's atmosphere at any point on its surface (caused by the weight of the column of air above it).

Atom The smallest part of a chemical element that can exist on its own. It has a central nucleus (made up of protons and neutrons), surrounded by electrons.

Bimetallic strip A strip consisting of two different metals bonded together. The metals expand by different amounts, so the whole strip bends when it is heated.

Boiling point The temperature at which a liquid changes into a gas or vapor.

Boyle's Law At constant temperature the pressure of a gas is inversely proportional to its volume.

calorie (cal) A unit of heat equal to the amount of heat needed to raise the temperature of 1 gram of water through 1°C.

Calorie (with a capital C) 1,000 calories, the same as a kilocalorie.

Capillarity Also called capillary action, the movement of a liquid up or down a narrow tube, caused by the attraction between its molecules and those of the tube. The surface of the liquid is curved into a meniscus.

Celsius scale A temperature scale that has 100 degrees between the freezing point of water (0°C) and the boiling point of water (100°C). It was called the centigrade scale in the past.

Centigrade scale An old name for the Celsius temperature scale.

Coefficient of expansion Also called expansivity, a measure of how much a substance expands when it is heated.

Coefficient of linear expansion Also called linear expansivity, a measure of how much the length of a solid object increases when it is heated.

Coefficient of volume expansion Also called volume expansivity, a measure of how much the volume of a substance (solid, liquid, or gas) increases when it is heated.

Compressor A machine for compressing a gas (putting a gas under high pressure).

Condensation The process by which a gas or vapor changes into a liquid.

Condenser An apparatus for converting a gas or vapor into a liquid.

Conduction The process by which heat moves through a solid object.

Convection The usual way in which heat moves through a fluid (a liquid or gas) by setting up convection currents in the fluid.

Crystal A solid that has a regular shape because of the ordered way in which its atoms or molecules are arranged.

Density For any substance, its mass divided by its volume.

Ductile Describing a metal that can easily be drawn out to form wire.

Electron A subatomic particle with a negative electric charge. Electrons surround the nucleus of an atom.

Fahrenheit scale Temperature scale with 180 degrees between the freezing point of water (32°F) and the boiling point of water (212°F).

Friction A force that prevents or slows down the movement of one surface against another surface.

Gas A state of matter in which the molecules move at random. A gas in a container takes on the size and shape of the container.

Geothermal energy A form of heat energy that comes from deep underground, as in geysers, hot springs, and volcanoes.

Heat The internal energy of an object that results from the vibrations of its particles (atoms or molecules).

Hydraulic press A machine that uses the pressure of a liquid to "magnify" a force.

Kinetic energy The energy an object possesses because it is moving.

Latent heat The heat taken in or given out when a substance undergoes a change of state.

Lift A force acting on a moving airfoil that keeps it in the air (because the lift is greater than the drag).

Malleable Describing a metal that can easily be beaten into a thin sheet.

Melting point The temperature at which a solid changes into a liquid. It is the same as the freezing point of the liquid.

Meniscus The curved shape of the surface of a liquid in a narrow tube, caused by capillarity.

Molecule A combination of at least two atoms that forms the smallest unit of a chemical element or compound.

Partial vacuum A region of low atmospheric pressure, especially one in which most of the air has been pumped out.

pascal The standard scientific unit of pressure.

Pressure The amount of force pressing on a particular area.

Solid A state of matter that keeps its own shape (unlike a gas or a liquid).

Surface tension An effect that makes a liquid appear to have a surface "skin."

Thermal conductivity A measure of the ability of a substance to conduct heat.

Thermal A vertical convection current in a gas, usually air.

Turbulence The irregular flow of a fluid moving around an object. It is reduced by streamlining.

Upthrust The apparent loss in weight of a floating object, equal to the buoyant force keeping it afloat.

Vapor Another name for the gas that forms when a liquid boils or evaporates.

Vaporization The change from liquid to gas or vapor on heating.

FURTHER RESEARCH

Books

Al-Khalili, Jim. *The Physics Book* (DK Big Ideas). New York, NY. Dorling Kindersley, 2023.

Holzner, Steve. *Physics for Dummies*. Hoboken, NJ. Wiley, 2021.

O'Daly, Anne. *The Basics of States of Matter.* New York, NY: Rosen Publishing, 2024.

Web Sites

Britannica Kids
https://kids.britannica.com/students/article/matter/275737
An introduction to states of matter, with videos, images, and links.

How Stuff Works–Physical Science
https://science.howstuffworks.com/physical-science-channel.htm
Topics on all aspects of physics.

Science Buddies
https://www.sciencebuddies.org/stem-activities/subjects/physics
Physics activities for Kids.

INDEX